Eric Wade built a cabin in remote Alaska where he took his young family, venturing hundreds of miles by small boat. They chased away bears, marveled at giant eagles, stalked moose, and discovered a greater understanding of family and nature.

"A wonderful, addictive love song to the Alaskan wilderness."—*Charles Rangeley-Wilson, author of Silver Shoals and The Silt Road*

"A poet with an axe, a teacher on a river, forever learning and sharing."—*Kim Heacox, author of Jimmy Bluefeather and The Only Kayak*

"A tale of decades spent learning, enjoying and sharing a rare gift."—*Howard Weaver, writer and editor at the Anchorage Daily News, where he worked on both of the paper's two Pulitzer Prize winning series*

"A soulful story of teacher turned student; a man bent on immersing himself in wilderness ways."—*Debra McKinney, author of Beyond the Bear*

"Belongs on the shelf of anyone contemplating finding their own version of the Alaska Dream."—*Tom Walker, author of Wild Shots: A Photographer's Life in Alaska and We Live in the Alaskan Bush*

Eric Wade found the perfect place in the vast wilderness of interior Alaska to move his family. He climbed the river bank to walk on the firm forest floor. He wove through the trees, brushed aside rose bushes, and kicked the ground like checking a tire. The land spread before him with majestic white spruce and views of a sparkling clearwater river. His family would grow to love the landscape as much as he did . . . but over time, his dream changed, as did the land itself.

Cabin: An Alaska Wilderness Dream

Eric Wade

Moonshine Cove Publishing, LLC
Abbeville, South Carolina U.S.A.
First Moonshine Cove edition November 2019

ISBN: 978-1-945181-719
Library of Congress PCN: 2019916013
Copyright 2019 by Eric Wade

All rights reserved. No part of this book may be reproduced in whole or in part without written permission from the publisher except by reviewers who may quote brief excerpts in connection with a review in a newspaper, magazine or electronic publication; nor may any part of this book be reproduced, stored in a retrieval system or transmitted in any form or by any means electronic, mechanical, photocopying, recording or any other means, without written permission from the publisher.
Book cover design by Grady Earls; interior design by Moonshine Cove staff. Cover photo by Doylanne Wade. Interior photos by the author and Doylanne Wade. Back cover painting by the author.

To my sons who hammered my screwdrivers into the lawn, and my wife who helped me find them.

Foreword

I am standing on the boat's bow, looking down through the water. The September sun warms my shoulders through a heavy shirt. Out beyond the long cast of my silhouette it sends back shimmering reflections: beds of emerald plant-life swaying in the current, mosaics of small stones worn smooth by the water, and the sand, like grated cinnamon, the darkest I have ever seen. Behind me in the boat are my three brothers. We have come to the confluence of two small rivers in search of Arctic grayling, the most powerful fighters of these northern streams. Somewhere down there beneath the surface the fish are finning steady in the current, themselves hunting for their evening meal. We think the largest of these will be in the deepest and swiftest flow, so I am standing with the anchor in hand, while one of my brothers is at the wheel, looking to position the boat where the main current of one river cuts into the high bank of the other.

Four fishermen feels like a crowd when you are on a boat built for small rivers. The best places to fish are from the bow and stern, so manning the anchor, or the wheel, is really a tactical move to put you in position for more elbow room when it comes time to cast. We all know this. Standing in the middle and casting to opposite sides involves some jostling for position, or, more tortuously, waiting your turn. The chance of getting whipped by fly line dramatically increases too, as does the possibility of getting snagged with a hook in some place that hopefully isn't your eye.

A fish rises to the first mosquito pattern placed on the water. Arctic grayling, especially the larger ones, rarely rocket out of the water when they strike, unlike, say, most species of trout. I see a gurgle on the surface, then the electric shock of line snapping taut, then the heavy, muscular pull as the grayling dives away. When I eventually haul it in there is excitement in the boat. Everyone stops to admire the gunmetal sheen of its side and its tremendous speckled dorsal fin. Any time we are fortunate enough to catch a grayling it is customary, at least in our family, to lift the dorsal to full flex in order to admire its colors. Photos are snapped, mental measurements taken, and the hook is gingerly removed from the fish's lip. Then it is back to casting for everyone, with the three who did not catch the first fish all the more impatient to land one.

Witnessing all this, from a high sandy beach on the bank opposite, are mom and dad. Our mom meanders down the beach collecting pretty rocks in an old tin coffee can, the can itself probably older than some of her children. Way out here, something as useful as a tin can would rarely get thrown away. The rocks are for her garden back home. Dad sits on a driftwood log sipping from a can of beer, a can that will also find some future use back at the cabin. He had cooled the beer by throwing the can into a mesh bag, tying the bag to the back of the boat, and letting it refrigerate in the ice-cold river water during the run upriver. He is clearly pleased to see the fish strike, but there is no way he would want to be amongst the hectic action on the boat, even if space would have allowed it. From time to time one of us in the boat holds up to him a fish we think is impressive, and he nods in agreement.

This whole scene could have occurred any time in the past twenty-seven years, from when our dad first took the entire family up to a cabin he built on a remote Alaska river. But it happened only this past fall. One cannot stand in the same river twice, because rivers change, and people change, but in remembering these kinds of moments it seems much easier to reflect on those aspects of ourselves and our experiences that stay the same.

In another sense, though, the wilderness can change us with astonishing speed and force. I think about it, now, as a kind of re-wilding. On the first day of a trip to the cabin my hands are soft, easily scratched, easily hurt. In just a few days they feel calloused over, hardened against things like rough spruce bark, wooden axe handles, coarse rope, and all sorts of prickly muskeg undergrowth that you might need to claw at to get up a river bank, around the edge of a swamp, or over a ridge. My knuckles roughen, my grip strength grows. At the start of a trip the river water will numb my hands in seconds, so that they feel fat and the fingers refuse to clinch. By the end I find bathing in the river almost (but not quite) comfortable. A bad mosquito bite will redden, puff up, and itch intolerably. By the end I hardly notice them.

Our senses sharpen too. In a world in which the only human-made noise is that which we make ourselves, we become attuned to the subtlest disturbances: the soft plop of a beaver slipping into a lake, or the whoosh of an otherwise silent owl drifting amongst the treetops. During hunting season my dad won't run a chainsaw for weeks in advance, for fear such an artificial noise might push the moose miles away. The tart scent of ripening cranberries in the fall, or the clean earthiness of a fresh-cut log are among the subtle smells of this northern land.

Sight, too, re-calibrates: shifts in light due to growing cloud cover and impending rain, the faintest signs of a game trail, the ghostly camouflage of a grayling against the riverbed. And food never tastes as good anywhere else. When we were kids, we would wolf down giant pilot crackers, sometimes, as a special treat, with peanut butter and jelly on top, but usually just plain. If any crackers happened to make it the whole trip without being eaten, they might as well have been thrown away. There was no way us kids would touch them in town. We always came back to civilization different, grown. Out in the wilderness we were shown a whole new way of living in the world, and it has changed the way we now live our lives day by day.

Some thirty years ago my dad staked land in remote Alaska. Then he built a cabin on it. These two unlikely feats, as the following pages show, are in themselves remarkable. But even more remarkable is that since then he, with my mom, have made it into a homestead. It has been something that binds our family together, that has afforded us unique shared experiences and, through them, unique bonds. The homestead is something that changed us, that still changes us, though it is also a fixed idea, as well as a location, that we can come back to. I am happy to say that my dad is someone who, through sweat and smarts and good fortune, along with no little appetite for adventure, has made good on his pursuit of an Alaska wilderness dream. Now he has written a book that chronicles his adventures, and I am happy about that too.

James Wade, Girton College, Cambridge

Author's Note

When I set out to write this book, time posed a problem for me. Not the time to write the book, although that's always a challenge, but rather how to organize time in the story. What comes first and then what? What order should the tales be shared? I also struggled with choosing which adventures were most worthy of telling. Thinking back on the thousands of river miles, all the escapades, and adding the perspectives of family and friends who ventured with me, I faced far too many stories to tell. So, I've done my best, selecting stories I thought most poignant, and presenting them in a nonlinear manner, to tell about my family's pursuit of wilderness adventures. So be advised, some time travel is required.

Cabin: An Alaska Wilderness Dream

1

Growing old on an old river

"Pursuing a dream can be a bad idea, I suppose, but sometimes you do something even if it's wrong." I was talking with my wife Doylanne. She pursed her lips and lowered her eyebrows, so they sloped inward. I'd seen that look before.

Her right cheek lifted.

"Eric, that's a silly philosophy." She stacked dripping dishes. "We can stay at the cabin for part of the year, but we can't move there, not for the whole year."

"Why? What's going on here that's so important?"

"Well, right now I'm doing the dishes. And you?"

"I'm supervising. By the way, you're pretty good at dishes."

During one of those cabin stays, many years later, in September 2016, with the sun easily overwhelming the morning chill, Doylanne and I, in our early sixties, left our wilderness cabin to see how far we could go up river. A rare combination of high water and extra fuel offered a chance to fish for grayling where I had never fished before. We'd been up river many times but never to where we could go no further. We ran by boat into the northside foothills of the Alaska range. The blue sky and yellow fall leaves offered Saturday morning cartoon colors to enjoy. I popped the boat onto step and slowed to cruise. For an hour we ran the winding river, drinking coffee. Every moment was astonishing. Moving at eighteen miles an hour, we wove up the channel until the entire opening, trees to trees, narrowed to the width of a house.

I cut the motor and veered to the bank. I felt it first, then saw it. It was a pathway to a rotting log cabin my brother Charles and I found thirty years before. The moose hooves that hung from a spruce branch were gone, as was most of the cabin. I'd aged a little better, I thought, but in fairness to the unpeeled logs, I'd spent most of my time warm and out of the rain. I've often thought about this remote cabin. Maybe this is what happened: A lone traveler journeyed up the shallow river one summer, not many years before my first visit, to build a cabin and take home a moose before freeze-up. His real purpose, though, was to find out who he was, maybe to prove to himself that he could finish something difficult. Using an axe, he built a cabin the way it was done in the old days. Only an axe and a file. He knew that he could separate himself from others by making things hard for himself. Building a cabin with an axe in the company of a trillion mosquitoes would do it.

Doylanne and I went on, twisting up river, and I hit the channel nicely and trimmed the outboard just right and mountain angles I'd never seen emerged. We stopped for lunch on a beach that crept out just down from a stream entering pale gold into the river. A place I'd never been. I pulled on chest waders and moved up the river for enough room to easily cast. Doylanne tidied the boat cabin and looked for pebbles on the beach.

I stood in the brisk current shin high and sopped up the place. "What is the strange gravity of a trout stream? Why do we slow our cars and crane our necks at every bridge? What exactly are we looking for?" Ted Leeson asks these questions in *The Habit of Rivers*.

Maybe we look for a time long gone.

I cast across the narrow river to the mouth of the stream, and before I could correct the drift of the line, a grayling struck. The grayling fought gallantly and came in calm and beautiful with only its tail moving like Hemingway's great fish. I released the fish. Doylanne filled a can with colorful little rocks. There's a feeling of freedom that comes from standing in a stream, floating in a boat, or gazing at river sparkles. I think it's because we are always moving on. We crane our necks to look back on a trout stream because we are leaving it behind and never returning. It's driving away from our hometown knowing we'll never be the same. Like a river, we never stop.

We ate lunch on the beach and talked about bringing up a canoe one day and paddling up the stream, but we admitted we might not ever see this place again. That was age talking. When you reach a certain age, of course different for all of us, dreams become what Raymond Carver called, ". . .what you wake up from." Maybe. We long for the first and the new, and that's a stream.

About five miles further up river, a creek, shallow and wide and the color of copper emerged from a field of high grass on one bank and a grove of birch on the other. I was hitting rocks frequently now, not stopping, but the skeg dragged the bottom and the prop slashed the rocks. The river was shrinking rapidly. I stopped above the creek, and we moved furtively down the edge of the grass to where the waterways met. A shoal of grayling visible across the stream held mostly steady looking upstream. I flipped a fly and a grayling shot for it.

I caught and released a few fish before we moved on up the river. Slow going. The current was unmistakable and easy to read. There just wasn't much water. At times the outboard prop clanked like a spoon in a blender.

We stopped when we could go no farther.

We would all crane our necks to look at this river.

I steered over to the bank, and with grizzly tracks as guides, we walked a hundred yards further up the bank, staying in view of the boat.

I thought about river currents, how our lives were shaped by sandbars, ridges, streams, ponds, and grayling on a fly. I measured a grizzly track with my hand and retrieved a rifle from the boat. Doylanne continued her search for the perfect pebble while I fished until dusk. Then we ran rapidly downstream, heading back to the cabin. I raced the light. I flicked on the headlights as I pulled up to the bank at the cabin and looked for a bear before tying off the boat and heading up the trail.

Only days before, we'd left our home in Wasilla for our annual fall trip. Doylanne was anxious to see the cabin in its new location. Three months before, in May, Jack, our oldest son, and I, using pulleys and a chainsaw winch, moved the cabin out of a pond of melted permafrost.

We benefited from high fall water levels so our journey was smooth without misadventure. The September foliage glowed. Chlorophyll dried revealing the yellow of xanthophylls and the orange pigments of carotenoids, colors of the autumn woods. Lovely place. We camped on a high sandy beach and watched the sun fade beyond our campfire. In the boat cabin, we slept like we were home.

I ran the boat aground once the next morning, but not bad, off in minutes. About one hundred and fifty miles in, Doylanne spotted movement on a starboard beach. I slowed, and it soon became clear a pack of wolves watched us. Three laid in the sand and two trotted in circles behind the others. I held steady in the current across from them, probably one hundred yards from the closest wolf. I was surprised by their light color, blending in with sand and sun-bleached driftwood. They all stood and stared momentarily then trotted up the beach. We watched them disappear in the brush.

We ran on and pushed against the bank at the homestead early in the evening. Doylanne ran up the bank before me. First time she ever beat me up the bank. I followed her with a rifle. She knelt.

"What's this?" she asked.

I looked for bears.

"Eric, look here." She cupped a rhubarb plant in her hands. Little one, but in September, a yellow, orange, red, rhubarb plant, fragile but there. "How'd it get here?" she asked.

I looked for water around the cabin. The water was behind us one hundred and fifty feet at the old cabin spot. There was no surface water at the new location. "Must have been from our boots in May," I said, but had no idea.

She stood and hugged me. "We made it."

Yes, we'd made it. Even when it's smooth running, two hundred miles of river is a hard trip. "Let's get in the cabin and call it a night," I said.

In the morning, we emptied the boat and drank coffee. I wanted to hike behind the cabin to see if there was a moose close before we scared if off with all of our cabin noise.

"Want to go? I'm going to walk out behind the cabin." A walk meant a few hours.

"No, I'll clean around here."

I headed downstream along the river a few feet behind the trees stepping over decaying logs looking for a bull moose. A light breeze blew from the south and Denali, rustling birch and willow leaves. The river sparkled. After about a half mile, the animal trail swerved to a small horseshoe lake where a pair of trumpeter swans paddled toward the reeds at the water's edge. Their white feathers and black beaks were stark against the gun barrel blue of the water and the olive-green trees. A squirrel ranted a few trees away. Out several feet in the marsh was a branchless tree. The top of the tree fell and flew away.

I turned ninety degrees left and shortly came to a meadow of tall brown grass and a chute to another small lake. Time to stop. A moose is a big deal on your back, but I could handle this distance. If I shot one here, I'd hang it in the woods and take my time over a couple days to carry it to the cabin.

I sat against a spruce, settled into my coat and knit hat, with the sun breaking over the trees, and fell asleep. I had slept many times before in the woods and once awakened to a moose. I curled on the ground, my face in the crook of my elbow. A whoosh brought up my head. A hawk swerved away at my movement. I felt warm enough to spring up out of my coil. It would soon be hot for September. An easy breeze moved the high brown grass, and hawks flew in figure eights around the meadow. No moose.

I often hear sounds that make me think a moose is near: crack in the woods, strain of a bending tree, slap of a wing, or squirrel bounding up a tree. I heard those sounds but didn't see a moose.

Time to go back to the cabin. After twenty minutes swinging to my left expecting to either run into the clearing at the homestead or the river, I stopped and ate a candy bar, confused about my location. I discovered the batteries in my GPS were dead. There it was, the eerie sensation of being lost. I was close to the cabin but didn't know which way to go. It unsettled me. I walked to a stand of tall spruce. No river but a drying horseshoe lake lie before me. I didn't recognize this place. A pair of swans slid across at the far end. I stopped, closed my eyes and listened. I heard the river, faint but true. A hawk landed on the headless tree. Guided by the sound of the river, I made it back to the cabin.

Doylanne and I worked steadily, enjoying a surge of purpose. The cabin, out of the pond, sat high and dry with beautiful views of the river and mountains. We found many things needed repair: fix the floor, install a new metal roof, add a new deck, and level the steps.

We worked on cabin projects and hunted and did it again and again. No moose. In the early afternoon of the last day of the season, we left for home in a light snow flurry that made it feel like winter. Close to dark, we saw a large black bear crossing a beach.

A couple turns further down river a bull moose stood alone on shore. Doylanne moved quickly to my seat — we'd done this before — and I went to the front of the boat with the rifle. She shut down the motor and steered toward the beach.

The moose stood face on at the edge of a large and dense clump of willows. It was the beach side of a river turn; a long expanse of pearl and oyster colored sand lay between us. A long shot for me anytime, but from a drifting boat, close to two hundred yards, sketchy. I fired. Seconds later the moose stumbled. I'd hit the moose, but it didn't fall.

Doylanne started the engine and ran toward the beach. The moose stepped back in the brush, still visible. The last of the leaves swirled with snow in the air. I jumped off the boat, fired, and the moose fell.

We worked most of the night on the huge moose, struggling with everything: Getting it in position to gut, skinning, cutting it into parts. Hours into working on the moose, exhausted and about to call it a night, I sawed the sternum and separated the rib cage, wedging the ribs apart with a nifty tool I received as a Christmas present. The wedge slipped. The top half of the rib cage slammed down on my right hand. Minutes later my hand was a bloody blown up latex glove. Two gray-headed sixty-somethings worked over a huge animal in the cold and dark. We carried on because we couldn't stop until we were done. The color-shifting sky morphed from black to silver then red to blue that morning before we trudged cold to the boat.

I coiled in my sleeping bag thinking about guardian angels and the signs I'd seen of them along the river through the years. Maybe I was too tired and not thinking straight. My hand ached. I pushed my thumbs into the base of my skull. It would be hours before my feet were comfortable. Doylanne fell off to sleep.

We lingered in our sleeping bags after we awoke, waiting for the other to get up and pee and start the heater. Doylanne knew this routine. She wasn't moving. I started the heater and the cabin warmed quickly. We shared coffee with cheese and crackers. A moose carcass was waiting to be loaded into the boat.

We enjoyed a special sight as we loaded the moose. A dozen spruce hen strutted across the beach. We were using the tarp-as-a-sled method to pull the quarters one hundred fifty yards from the trees to the boat. The capstan winch pulled the rope taut just inches above the sand. The spruce hen ran to the rope and hopped over it. They stayed with us all day jumping rope.

It snowed lightly all day, the flakes blowing away with brief gusts from the north. In fading daylight, we pulled away from the beach to make some time down river. I powered the boat onto step but slowed at every crossing. The biggest moose I ever bagged filled the front of the boat. Two hours later, with the sky constantly changing costumes, we called

it a night. We shivered, but our small propane heater kept the boat cabin tolerable, and we lit a lantern. We settled in slowly as the boat cabin warmed, talking about how heavy we were and the shallow corners we knew we would encounter down river.

The chorus howls of wolves awoke us during the night. How far away would they be to sound so close? We talked about wolves as we hid from the cold in our bags. I thought the wolves were coming closer but kept that to myself. Doylanne shined the flashlight at the aluminum ceiling covered with frozen droplets of condensation.

"We should scrape those off now. We're going to get wet tomorrow," she said.

"You could be right. Want me to hold the flashlight?"

She shoved me and turned off the light.

The wolves howled again. They were closer. "The gun's right here," I reminded Doylanne.

"Is it loaded?" She couldn't see me, but that made me smile.

"When someone shows you who they are, believe them the first time," wrote Maya Angelou. I was right to believe Doylanne when she showed me who she was forty-eight years earlier.

In the morning, the near frozen fog rested on the water and visibility ended within feet, so we stayed in our sleeping bags. Several hours later the fog cleared in the cabin. I made coffee and Doylanne put together a cold breakfast, and we waited in a gray world. By late afternoon I could see enough river to putt, so we left. We spent another night on the river in the fog.

That winter I read *Travels with Charley*. Steinbeck wrote:

> When I was very young and the urge to be someplace else was on me, I was assured by mature people that maturity would cure this itch. When years described me as mature, the remedy prescribed was middle age. In middle age I was assured that greater age would calm my fever and now that

I am fifty-eight perhaps senility will do the job. Nothing has worked.

Back at home, Doylanne and I waited for the river ice to thaw.

2

Down the meandering pathway

We're water drops on a floating leaf. A breeze, a twig, a rising fish can easily topple such a craft. Go anyway, go slow. Follow your dream.

Doylanne's cancer diagnoses that next winter startled us. Startled her doctor too. Doylanne looked in perfect health. I'd known her a long time and learned something new about her during this ordeal. She stepped forward, demanded information and access, and became her best advocate. The surgery went well the doctors said. They removed all the cancer and her recovery seemed to go as hoped, so we were excited about a family reunion and a niece's wedding in Arizona, but days before the trip south, Doylanne was admitted to the hospital through the emergency room, seriously ill with a septic blood infection. We cancelled our trip. In February, Doylanne went with me to a work meeting in Washington D.C. She looked around the city and read while I attended meetings. Our third day there, she was admitted to an Alexandria hospital with a blood infection. She was there nearly a week gravely ill. We went home and watched and worried.

"We should stay home this year," I said.
"No, I want to go."
"You won't be able to get to a hospital in a hurry."
"Not if you run aground."
"No, really. You know how that would go."
"I'm fine, Eric."

We pulled out for the homestead as Hurricane Harvey hit Texas in 2017. Rocky, our fifteen-year-old white toy poodle, rode with us. Water was the lowest I'd seen it in late August. We were heavy as always and cruised about twenty miles per hour into an uncertain sky until the outboard started knocking. I knew it was bad. I stopped and pulled into a small enclave. Stone gray evening on the river, we readied the cabin for the night, working in silence. This was obviously not what we wanted. From the boat cabin, we watched a handful of boats dash by headed to hunting camps.

We arose early to cold and headed back to the landing. The lower unit clanked for ten minutes. A snap and whine, the gears were gone. I steered and directed with the pole until we hit sand on the shallow side. We were now somewhere near twenty miles downriver from the landing, still a long way to go.

In the boat was a thirty horsepower outboard bought in 1988. I stood in the river, Doylanne in the boat. We lifted the motor to the lip of the gunwale where we rested. I pulled the motor over the side and stood it upright in the water and soft sand. One hundred twenty plus pounds in water and soft sand is serious business for those of us who just turned sixty-four. I wrestled the motor down the length of the boat. Pick it up, take a few steps and rest. On the transom was a one fifteen four-stroke outboard that I wasn't moving, so I had to put the thirty beside the one fifteen. I'd measured it, and it would just fit. I lifted and guided the outboard. On the transom or in the river. I teetered at my physical edge. It was close, but the transom won. I mixed two stroke gasoline, and we were off upstream at two miles per hour. On one stretch, I turned into a chute that should have been a shortcut, but it slowed us to about a half mile per hour. The GPS kept track. Eleven hours later we reached the launch.

Eleven hours sounds bad, and it certainly could have been, but the weather was nice, and we had a satellite phone that kept communication alive. Our son Jed bought a new lower unit in Anchorage, and his wife, Carly, delivered it to our son Jack in Wasilla. Doylanne drove down the highway to cut a little into Jack's drive. They met at the road entrance to Denali Park.

Beautiful Rocky, three-pound blond toy poodle, bone skinny with weepy eyes, could barely stand, but we had to take him on this trip, couldn't leave him in his final days. He stayed with me and the boat when Doylanne drove down the highway to meet Jack. Rocky stayed in his kennel breathing deeply while I talked to him.

I installed the lower unit on the beach in the morning. While I worked on the motor, a guy came by to see what I was up to and ask where we were going. That's always awkward for me. Like a fishing hole, I don't want to disclose the location.

We pulled out again. Nice sound. High end four-stroke motors purr like high-end car engines, which is what they are.

At mile seventy, a thunderstorm forced us to beach the boat and get far away from the gasoline. Lightning was striking near us, and thunder shook the boat. Doylanne and I dashed into the trees. The rain turned to hail. It pelted us, and we knew there was no middle place here. If lightning struck the boat, Rocky would be dead but so would we. If we survived the explosion, fifty feet away, a wet and cold Alaskan night would probably kill us. We stood in a clump of willows and became soaked. We talked and shivered until the storm passed. We ran back to the boat and checked on Rocky and changed clothes in the boat as the clouds swept overhead and the sky re-focused to blue. A few hours later, we witnessed an amazing display of northern lights. We laid in our sleeping bags looking at the sky.

"Doylanne, you okay?" She was in long underwear and a knit hat, her nose poking out of her bag.

"I feel good. Thanks."

Our river journeys through the years drew us closer to one another.

The next day, after several running hours, we stopped at a walkway leading up a steep bank, a place we had stopped nearly every trip up the river through the years. We stopped to visit friends, a trapper we'd known for decades who always encouraged us along. He was now in his mid-seventies, and he and his lovely wife had built a new lodge style log house for a home. Tough, kind, and wonderful people. After a short visit, we continued toward the mountains, reaching the cabin near dark. It had been an odyssey for Rocky. He didn't want to come out of his kennel. Doylanne wrapped him in a blanket and put him in a flannel pillow case. He slept between us in his own sleeping bag that night. In the morning, he looked at us but wouldn't move, his tongue drooping, only three teeth left. Doylanne washed him and placed him back in his sleeping bag, palliative care for Rocky.

The meandering pathway to the outhouse with its hatchet door handle is one hundred and eleven steps along the river from the cabin, far enough I get lost in thought along the way. I counted the steps after reading Robert Macfarlane's book about paths, *The Old Ways, A Journey on Foot*. He explains the language relationship between walking and thinking:

> The trail begins with our verb to learn, meaning 'to acquire knowledge'. Moving backwards in language time, we reach the Old English leornian, 'to get knowledge, to be cultivated'. From leorian the path leads further back, into the fricative thickets of Proto-Germanic, and to the word liznojan, which has a base sense of 'to follow or to find a track' (from the Proto-Indo-European prefix leis-, meaning 'track'). To learn' therefore means at root — at route — 'to follow a track'.

We think when we walk. We learn on our way. The longer the path the more we learn.

In September, the path to the outhouse is always covered in red and yellow leaves of birch and highbush cranberries. About midway is where we buried Rocky. I felt his last breath as I sat by the woodstove and watched the river roll to the sea. Rocky walked a long path, from South Dakota to Chicago to Wasilla to the homestead. His partner Daisy passed the year before.

We don't know what lies in store for us, but we know that change is required. I've never passed the ten-year test. When I was fifteen, I was a high school sophomore in Oregon; when I was twenty-five, I was a school teacher in rural Alaska; when I was thirty-five, I was a school principal in Alaska with four children and on my way to stake a homestead in interior Alaska; at forty-five, I worked for an outdoor education association; at fifty-five, I worked for a conservation district; at sixty-five I work with families who have children experiencing disabilities. I could have never guessed all those changes.

It's not possible to predict the outcome of a journey. Doylanne says the outhouse path has become shorter over time. I agree. It's easier to maneuver, the route more predictable, but the experience is never the same and the outcome never certain. Each trip down the short path is different, every footfall special, and we learn. The outcome is not known but we move on because we must step one more time.

Two guys pulled up to the cabin the day after we buried Rocky. We talked about pets and growing old and dying. We all agreed Rocky had a nice spot on the river. I guided them around the cabin, and we inspected bear claw marks. The talk soon turned to moose hunting. They were both from the Midwest, one of whom moved up a few years before. The other was visiting to get a moose with a bow. They'd come a long way. Talkative and friendly, they wanted to hear stories about big moose. They'd come this far into nowhere, to the middle of Alaska; there must be bulls around. The bow hunter would be calling one in for a shot.

I told them they had a chance, but moose were hard to see in this brushy country. They left putting upriver. They looked for a big bull, they said, a bull with massive antlers.

Bull moose undergo an amazing transformation every year. A bull drops its antlers during the winter to save energy and grows them back in the summer. Antlers can be huge and often weigh more than sixty pounds. In this area of the state, in mid-September, moose have already shed the velvet from their antlers by knocking and scraping them against brush and trees.

Doylanne and I wondered about the hunters, worried about them. Five days later, they motored slowly by the cabin. A blue-bird day, unusually warm, Doylanne and I were on the deck painting window trim. They were heading out, they said, clearly dejected. They hadn't seen or heard anything. They told us this while drifting by. They didn't want to stop. I suspect someone had told them this was the place to go for big moose. It might have been one day. I certainly believe more moose lived around here thirty years ago. I felt for the hunters, but you can't call in a moose if it's not there. Moose are moving to higher country and drier ground. The habitat is changing.

I walked out back of the cabin, circled a mile or so and became dismayed by all the leaning, drunken trees, many of them big trees, birch and spruce that once grew straight but now leaned together, holding each other up. Like moose, trees are finding it hard to live here.

Those big moose hid from the hunters who visited our cabin. They hid from me too.

After they left, I started hunting with some purpose. The low water kept me close to the cabin. One lower unit died already, and two hundred miles of river flowed to the launch. Doylanne and I checked the river level several times a day. The sun shone here, but maybe rain lashed the mountains and the waters would rise. We hoped. We'd never seen the water like this before in late September. Should we go? We stayed. Five days of hunting season remained, and we would miss the place over the long winter.

One morning, I sat reading a story about medieval fairies written by James at a spruce top table made by Jack, wood taken from a fire-killed spruce cut two hundred feet from where I sat. I spun slowly in the cabin inspecting the cabin's spruce logs, all taken from the yard. The white spruce logs had darkened over the years. I'd never put a finish on the inside logs so the deep brown leather appearance resulted from cooking oils and vapors, exhaust from gasoline lanterns, smoke from candles and the wood stove, and I suppose just us. The beautiful logs used to build this small cabin contained the board feet to build a twelve hundred square foot house. I promised myself, if I built again, I would mill boards and save some of the trees.

A spruce grouse landed in a birch in my view. We looked out all the windows. Grouse encircled the cabin, nine we could see. We imagined the bird in the tree watched for the others as they had their breakfast. Spruce grouse are the lambs of the wilderness. We watched, delighted by their awkwardness. No need to bother them. They walked circles around the cabin, and we drank coffee.

An hour later I headed down the path to the outhouse, head down, book in hand. A noise brought my head up, and there stood a bull moose. One hundred feet away. I didn't have a rifle. The bull, with a forty inch or so rack, ran to the woods. A moose walked in the yard and the hunter lacked a gun. Classic. We were always pounding on something and chopping wood. The moose likely responded to our calls.

I guess I don't care enough about bagging a moose.

In late afternoon, Doylanne and I headed slowly down river to a meadow where I'd shot moose before. I tilted the motor as high as possible to avoid the bottom and logs. The greenish surface disguised water as clear as cleaned glass. The channel, mostly visible, zig-zagged down the straight stretches and hung to the inside corners, but there wasn't enough water in the river to make the channels easily passable. We bounced off logs at five miles per hour for about three miles. Early in the trip an eagle kept track, but it must have become bored. It moved on and was replaced by three swans circling before dropping behind some spruce to a horseshoe lake where we sometimes picked cranberries. We heard them glide in for a splash landing.

    I tied off to a small birch, the sun hitting the bow directly. Sandhill cranes passed overhead. A long beach with moose tracks stretched upstream on our left and a long expanse of meadow to the right. We sat in the bow in the sun. Soon I took my shirt off for some subarctic sunbathing. Facing the sun, my chest was warm and my back cool. I suggested Doylanne take hers off too. I tried again a couple times. She suggested I focus more on hunting. Warm, beautiful afternoon, I grabbed my fly rod and after a couple casts caught and kept a two-pound grayling. Near dark, we putted back dragging sand and bumping logs. Ten-dollar solar lights lit the cabin yard. That night a bear stood on the deck.

    One of Doylanne's projects during this trip was sprucing up a table Jake cut his face on years before. She sanded some on the plywood top, bleached it and brushed on several coats of polyurethane. Lots can be done with a piece of half inch plywood. It's now a beauty and will hold coffee cups on the deck.

    After a night of steady rain, the river rose three inches, not much but the right direction. Time for us to head back to town. We'd been gone twenty-five days. Doylanne and I fought two barrels of fuel into the boat, one fifty-five-gallon and one thirty-gallon. I'd winched them onto shore when we arrived. The low water kept us from using all the gasoline, so we carried a fifteen-gallon drum to the shed. I lifted one side of a board run through the top handle of the gasoline drum and she lifted the other. Doylanne's now as strong as me, maybe stronger. I rested first. She said my half was heavier than her half. Yes, of course it was.

More than usual at this point of the trip, I grew sad. For some reason it's now easier to grow sad. With some copper foil brought in to cover a countertop, I made a grave marker. I wrote Rocky with wire and placed the letters on a board, pounding the copper over the letters. Doylanne added some river rocks at the base of the spruce board marker and threw Shasta daisy seeds on his grave, hoping for flowers in the spring. For the first time since Jack was born, thirty-eight years before, nothing depended on our care. It felt like a long winter ahead.

The evening before we left for home, we sat on a meadow's edge listening to the earth and waiting for a moose to appear. The earth occasionally held its breath, and the woods became as silent as the moments after a chime. We listened like we do when standing over a baby's crib. The music of the earth drained any desire to kill a moose. We held hands that evening, sitting in a cranberry patch, until dark forced us back to the cabin.

 We pulled out for home with plans to spend a couple nights looking for northern lights, cranberries, and moose. A lot can happen in two hundred miles of river; there could still be a moose out there for us. We picked our way through the shallows an hour below the cabin (what should have been an hour trip took four) and ran into some hunters who failed to make it up through them. We exchanged stories for a while, and they talked about watching a couple guys stuck for most of a day on a sandbar, but they couldn't get close enough to them to help. They were probably our visitors from the Midwest.

 We navigated down river in low but doable water. We were tired but rejuvenated by the cold and the beauty. Doylanne felt well, and we cruised slowly along. We camped where it looked like a moose hangout, but darkness came too quickly for hunting. We slept in the boat just a little cold and a little uncomfortable and loved every minute of it. We ran on the next day, but it took another day of fighting the water and another night sleeping in the cold and listening to river sounds before we reached the boat launch.

3

Land of Wooly Mammoths

I dreamed I watched the river through quivering birch leaves. I sat on a hillside above the silver water while wind gusts sustained sand twisters on eggshell colored beaches. A butterfly landed on my knee and sat with me for several minutes before fluttering off to a cranberry bud where it bloomed into an angel.
"I saw you on the hillside," I said.
I awoke to roll to my side, struggling to breathe, pain radiating across my chest.
"We have to go, Dad. They're here," My sons Jack and Jed ran out of the cabin, and I lay under a hissing gasoline lantern. I heard their snow machines start and fade into the trees.
Two days before, we left for our cabin on snow machines, departing from the highway hundreds of miles north of Anchorage. We left at daylight on a trail that petered out at about twenty miles. Jack led the way navigating by GPS. He broke trail through deep snow, and Jed and I followed pulling sleds with gear and gasoline. We inched along, and by dark we were at a one-room abandoned plywood structure on an unnamed lake about half way. A barrel wood stove and a couple bunks stood intact, but the windows were only openings and the door lay near the lake. We covered the windows with blankets, put the door in place, swept out the cabin and started a fire. The cabin warmed but was still frozen near the floor. Outside it was fifteen below. That night Jed slept cold, his sleeping bag not quite warm enough.

He coiled around the stove. Jack and I reached the bunks first. Jed could have joined one of us, of course, there was room on the bunks, but chose to make it where he was.

In the morning, we faced a decision: follow a natural corridor between the hills and linking lakes and follow the river to the cabin or take a shorter, straight line route through the forest. Both would be breaking trail we figured. But we were wrong. One way was breaking a trail, the other making a trail. We left straight for the cabin through the woods, taking the shorter way, making trail.

For a long day we crawled through dense brush, through forest fire debris, and across sunken creek beds. In places, the ravines dropped radically, and we lowered the machines with ropes and lifted them up the other side with a come-a-long. When it turned dark, about 7:00 p.m., we kept going. We knew we were close to the cabin. On one tight turn, I tipped over my machine and Jed ran over me. One of my legs trapped under his machine, the other across his lap. No injury but we all had enough.

"Dad, we've got to stop," Jed said.

Bitter cold, we dug into chores to get a camp set. At 11:00 p.m. I crawled into my sleeping bag in the tent. I shivered and twisted but couldn't shake the chill. The boys came in after me, and I heard them turn restlessly in their bags before growing mostly silent. My sleeping bag wasn't good enough that night. I tried to create space to avoid the bag itself.

In the morning, I awoke to sounds of Jed and Jack talking outside the tent and the snapping of a campfire. During the night, I rolled off my sleeping pad and my bag froze to the tent floor. I pried loose, and in a sore blue morning light, crept out of the bag and struggled to get my feet in my boots, their tops frozen perpendicular to the soles. I drove cold feet into frozen boots.

"Dad, look there," Jed said. It was the river, and I recognized the corner. We were camped on the edge of a high bank five minutes from the cabin. We lowered the machines with ropes and headed up river.

The cabin was magnificent in the snow. We quickly started a fire in the woodstove. Breakfast and lots of coffee. We were so excited about making it to the cabin, we skipped the nap we'd planned and went exploring instead. The day warmed to twenty degrees, and we covered miles of snow and ice. When the sky finally turned pink, we headed to the cabin.

I went to bed and fell to sleep listening to the boys talking about moose hunting and future trips. I'd worked myself too hard and felt sick. At 2:00 a.m., I sat upright in bed then I fell back squirming. I called out to the boys. They sprung up starting the lantern. Heart attack. I knew it. I writhed in bed, a wooly mammoth on my chest.

Jack called 911 on the satellite phone and passed on the coordinates. From the middle of Alaska, Jack connected with civilization. I heard him talk and describe the situation. I stood and raised my arms above my head and fell back on the bed.

"They'll get back to us," Jack said. "I'm on hold."

They were right back.

"Dad, they're coming."

We were too far out for anyone but the military. The Alaska State Troopers contacted the Alaska Air National Guard who rapidly dispatched an HH-60 Pave Hawk helicopter and HC-130 Hercules aircraft from Anchorage.

I rolled frantically on the bed.

Facing high winds and a blizzard, the helicopter turned back before reaching the Alaska Range, but the plane rose above the turmoil and kept coming. Jack stayed on the phone and relayed messages.

I struggled to breathe. I heard the aircraft blaze over the cabin about three hours after the initial call.

"We have to go, Dad," Jack said.

Jack and Jed drove the snow machines to a clearing about a half mile behind the cabin and shined their lights across the opening. I heard the HC-130 Hercules aircraft circle above us. Two pararescue men from the 212th Rescue Squadron jumped into the dark and the storm. I was on the floor when the pararescue men started working on me. EKG, morphine, monitoring, I was in bad shape. I knew it by their talk:

"See that?"

"We can't give more morphine," one of them said after my third dose.

* * *

A wooly mammoth makes a hissing sound and from a distance looks like a lantern. The mammoth drank from a pond as cave hyenas dashed across the meadow. I looked for moose in a stunning expanse of tall amber grass. A light breeze subdued the mosquitoes. The colors of autumn dazzled, and the mountains stood sharp against the ultramarine sky. A ridge angled downward to my left to a notch. I wanted to be up there, and at the same time, I didn't. A hawk swirled near me and dove into the grass. I heard them first, the bugle calls of Sandhill cranes. The rattles grew louder, and there they suddenly were, perhaps a hundred of them in v-formations, wave after wave passing across the sky. When the cranes disappeared, and their calls faded, I heard leaves in the wind, a faint fluttering, whispering sound, like a child sleeping. I hiked to another slight rise for another vantage point. I knew a moose hid in the grass. I called for one to stand. Nothing. Quiet and alone. We seldom have chances to be alone. I watched the hawks swirl and dive and thought of the crazy world. How come we don't like one another? Doylanne and I were in senior English class together. I carried her books to class. Our teacher liked me but didn't like Doylanne. She was too nice. He liked me, I believe, because he thought I wasn't. He would say to me, "Mr. Wade, do you have feet of clay?" He taught that kids and adults are mostly the same and not to get our hopes up. He also taught about fallacies. One of many fallacies he exposed us to is the error of attacking a speaker's argument by attacking the speaker. This is not childish behavior, just sounds like it. The people who practice the *Ad Hominem* technique are hateful. They create an ugly, distorted noise and know exactly what they're doing. We cover our ears. We pick out the parts of the conversation we are familiar with and too often reject the rest. It's hard to listen when the noise is grating and relentless. It makes us mean. Unplug the noise machines. There, again, another wave of cranes flew across the face of the great mountain. A saber-tooth tiger watched me cross the open meadow.

* * *

"Let's get him in the stretcher."

About 8:00 a.m. an UH-60 helicopter with two pilots, a crew chief, and medic from Fairbank's Fort Wainwright 16th Combat Aviation Brigade hovered over the river. The crew and the boys carried me to the helicopter. As they fastened me in the helicopter a young crewman, bless his heart, sat near me asking about fishing.

The helicopter lifted and disappeared over the trees, leaving Jack and Jed. Years later, Jack described the feeling after the helicopter left as an ache, a deep hip pain with no hope for respite, a shake your head, oh man, kind of pain. There was so much unknown. In biting temperatures, they pulled my snow machine into the cabin and left for the highway. Both of their machines struggled, forcing frequent fiddling to keep them going. They spent the night in the cabin without windows. The next morning, about twenty miles from the highway, Jack's machine quit. They ran double then, agonizingly slow. They curled their fingers in their gloves and froze. A week later, son Jake and friend Jeremy ran in and rescued the abandoned snow machine.

It took a few stents. The doctor at Fairbanks Memorial Hospital said my heart took a whipping, and it probably wasn't my first heart attack. I'd be weak for a while, he told us. Doylanne was with me by then, having flown up from Anchorage. We held hands and talked. Jake and Jeremy drove up. I was a fortunate man.

I recovered steadily that spring. I blended slowly with work. The days lengthened, and the sun hung longer at the broken tops of the leafless birch near the house. Often, particularly during the slow moments, when the rotting snow berms fought to hang on, I thought about the cabin and worried about being away from it. I watched the sky and thought of satellites and space stations and how satellite phones work and imagined what the cabin region of hills, rivers, and trees would look like from space. Maybe like this: a dark earth with Fairbanks, the nearest city, illuminated by lights and the largest lakes a white reflection when turned toward the moon. The cabin, just a little more than two hundred miles away from the international space station passing overhead, would be hidden in the expanse of forest. I thought about the cabin and the grayling circling in the deep holes in the frozen river.

As May came on with the first leaves, the red-breasted robins appeared, hopping across the greening spring lawn. They knew Doylanne and I sat on the deck. I could feel them as they hopped in the dew. I felt much better. We'd be going back to the cabin soon.

4

Where wolves and bears roam

Off a small river a log cabin stands with a handful of small rustic wooden structures, all concealed by willows, birch, and a grove of towering white spruce. A green metal cabin roof, beaten by the wind, snow, and scrapes from falling branches and talon marks from owls and bald eagles slopes gently, and over time the force of the snowpack has molded the roof line to an uneven shape, as if drawn in freehand. White spruce log walls rough to the touch, scarred by a chainsaw and the occasional bear claw enclose a rudimentary one-room wilderness cabin. Amongst the shrubs and grasses, the cabin looks like the forest, bent this way and that, pressed into shape by the elements.

How did we find our way to the Alaskan wilderness? For more than thirty years, I've traveled north of Denali to the cabin. During those journeys, I boated the same rivers, encountered the same corners, watched familiar trees grow and became well-acquainted with a place and a couple hundred miles of river. I went there the first time in 1987 to stake a homestead with plans to quit my job and move my family to the wilderness. Since that first trip, I've motored to and from the cabin more than seventy times, a distance equaling a circle of the earth.

On soft summer days, the wind blows from the south, swirling gently over the brush and around the trees and settles where it wants. The wind dictates the smell. With a breeze, the smells come from the river: mud, fish and wet grasses. No wind, the landscape smells like a spice cupboard with spruce needles, rosebuds, and birch bark. Often in the late afternoon, I catch the scent before a thunderstorm; and sometimes, when I'm most attentive, a special sweet smell of floral perfume.

I knew I'd found the perfect cabin location when I first climbed up the river bank to walk on the firm forest floor that first trip. I wove through the trees, brushed aside rose bushes and kicked the ground surface like checking a tire, startling a rabbit that darted away a dozen feet ahead. The land spread before me with majestic white spruce and views of a sparkling clearwater river.

The place occupies my mind every day and has for decades. Thoughts, photos, questions, discussions, worries, memories, all present as I meander through my day. When I reflect on the landscape where this story takes place, delicate and beautiful like the rose that dominates its forests in June, I realize I thought of it as a fixed place, resistant to great changes, but now I see a place where small variations make big differences.

***

My family was with me on a warm July evening in 1992, several years after my first trip to the property. Close to midnight, an owl's call silenced a twitching, chattering red squirrel. The squirrel, darting and leaping among spruce branches, watched me take a shower at a back corner of the cabin.

I stood below a black plastic container filled with water hanging from a rear corner eave. I heard a step behind me and turned to face a black bear standing twenty feet away. Naked and wet, I scrambled for my clothes, yelling, "There's a bear in the yard. Doylanne, stay where you are. Jack, get the rifle." Our sons were in the cabin.

The bear reared for a moment, taller than me, dropped to all fours and sprung over cranberries toward the blue polyethylene tarp outhouse near the river. Our small log outhouse flooded that summer, so I was building a new one on higher ground. Doylanne screamed from the temporary outhouse as the bear moved toward her. Jack checked the safety on the rifle then handed it to me while I tried to keep my eyes on the bear. Doylanne stuck her head around the corner of the tarp.

"You want me to stay here?"

I was trying to find the bear in the brush.

"Sprint! Don't stop!"

Doylanne, in a nightie and tennis shoes, came running, long legs high-stepping through cranberry bushes, dogwood flowers, and bluebells, over squirrel-made spruce cone mounds, and near a small stack of spruce firewood the boys gathered. She flew to James who caught her on the pathway near the cabin deck.

The bear emerged nose up where Doylanne departed and stopped to watch. It waddled out of sight. We all assembled on the deck where Doylanne counted heads as we watched shadows move in the yard.

"Is that it?" Jed asked, pointing to a root wad, partially obscured by a white spruce and lighted by the moon.

"No, there it is," James said. The bear plodded across the yard. The boys began throwing split firewood and screaming at it. The bear ran out of sight but soon reappeared and moved to within a few feet of us. This time the boys hit the bear with the firewood, and it jumped, bounded back, but didn't leave. I retreated inside the cabin with everyone behind me. We blocked the cabin light. The show was about to begin. The rifle barrel extended out the doorway.

"Dad, you've got to kill the bear." It was James.

"Cover your ears." The high-caliber rifle roared. The bear jumped back and stopped. I shot high to miss. It circled the yard. It wasn't going anywhere. The bear soon stood within feet of the deck, illuminated by lantern light escaping the cabin. I pulled the trigger again. The bear crashed back against a birch, biting at its chest as it rolled and stopped at the edge of the light. Before I could reload the rifle, it bounded into the brush.

Damn it.

I bolted the cabin door. We sat and listened to the wailing bear until James described Mom running across the yard. That lightened the mood a little, so Doylanne made popcorn.

* * *

Whiskey, philosophy, and bears mix together well. Several years later, early 2000s, Charles and a friend were at the cabin with me. We had a few drinks while watching the river and reliving outdoor adventures.

"You're supposed to fight back if attacked by a black bear," our friend said. "Hit 'em in the nose. That's the best defense if you're on your back."

"I'd certainly use a stick if I had one," I said.

"If you're on your back, you're dinner," Charles said.

"No, Chuck, not necessarily, it's bear psychology," our friend said. "See, bears don't know what we are. They don't know what we can do. A black bear isn't much bigger than I am. It sees me, it's going to think twice before taking me on."

Charles and I shared a glance.

"Here, take another swig," I said.

Our friend went on, "A bear is about six-foot tall on its hind legs and weighs three hundred pounds or so. Not that much bigger than me. I'd definitely fight one if I had to."

I passed the fifth of whiskey to Charles. We all stared off for a bit.

"Bears have big sharp teeth and claws. Let's not test it," Charles said. "I'm going to bed." Charles took off for the cabin, picking his way along the edge of brackish water from melted permafrost.

It was going to be an evening with deep snoring. It took us a while to get settled into our bunks, but minutes later our friend let out a yelp. The head of his bed aligned with a window and his nose was a fragile pane of glass away from a black bear. We scrambled to windows to see what was going on in the yard. The bear waddled toward the permafrost refrigerator. Say goodbye to our food, I told them. I left my gun alone and grabbed my camera. We were about to lose our butter, cheese, eggs, and steaks.

Charles's friend, though, went temporarily insane. In his shorts and t-shirt, he charged off the deck and ran screaming directly at the bear. He covered about thirty yards at full speed, arms waving, and screaming. The bear turned toward him, but only for a moment before it scrambled up the nearest spruce tree. In a moment of generosity and serious lapse of judgment, or advanced bear psychology, our friend saved the food and achieved legendary status. Charles and I cheered him on. Back in the cabin, we watched the bear sniff the refrigerator and wander through the yard before disappearing in the brush.

The bear ten years before wasn't so lucky.

* * *

Jack, James and I, all of us smelling like a movie theater, popcorn on our breath and butter on our sleeves, crept to the edge of the clearing. We peered into the thicket, a moon streak at our back. Jack carried a shotgun and James a lantern. When I stepped forward into the brush with a rifle, they followed. The hissing gasoline lantern carried all the light. Black beyond ten feet, we stepped forward again. The bear moaned. In the yellow light of the lantern, we stood motionless and listened, until I shuddered. Enough. We sprinted back to the cabin. We talked until everyone fell to sleep. Not me, though. No sleep. In my mind, over and over, I played a frightening movie. I should have known better than going after a bear in the dark. With Jack and James to boot? Was I nuts?

Jack found the bear dead the next day. He followed blood for more than a quarter mile northward to a cranberry patch. Entrails guided his way the final few steps. We tied the bear's paws together and ran a spruce pole between them but couldn't lift it, so we gutted it there. We then carried the bear to the yard where we skinned it, hung the hide between two trees, and worked on it for hours. We all took turns scraping fat from the hide. There was no way to keep the meat fresh, and it wasn't legally required, so we pulled the carcass to the boat and ran it downstream and pulled it into the woods.

In the evening, just at dark, the river's edge is a good place to stand with the roses and high bush cranberries and listen as swans honk and slap their wings on the nearby lakes. I often stroll there as the cabin settles for the evening. Three nights after I killed the bear, wolves sang. The skies moaned, and I imagined heads turning throughout the woods. The cabin hummed with a noisy gas lantern and chattering children, but Doylanne heard the howls. She came to the deck. Cold night, breath visible, we waved at each other and listened.

Later that night, lying in the one room cabin with the murmurs of sleeping children, I listened for the sounds of blood coursing through my body. I thought I could hear it. I turned to my side in bed trying not to make any sound. There was a steady thumping in my ear against my pillow, and beyond the log walls, the noises of the wilderness. No chattering squirrels, but the wolves were still there, soulfully calling between a gut pile in a cranberry patch and a decaying, skinned bear down river.

No moose either. I listened for one, but that would be a very silly moose. They would be hiding silently in the thickets. In late September in the north woods, there are two sounds that tear at your heart: Howling wolves, certainly, and the lonesome calls of moose, the unique creatures Thoreau called "God's own horses."

I listened and listened and eventually fell asleep.

5

Somewhere between Denali and the Yukon River

Concerning the factors of silence, solitude and darkness, we can only say that they are actually elements in the production of the infantile anxiety from which the majority of human beings have never become quite free. —*Sigmund Freud*

Since boyhood, I've wanted to live in the wilderness. I fancied a cabin, an ax and a notch, a flannel shirt, a wood stove, a cast iron skillet, and a cache in the trees, and always, for as long as I can remember, a place far, far away. James Fenimore Cooper's *Last of the Mohicans* is partly to blame, as well as frontier woodsmen like Daniel Boone, Davie Crocket, Kit Carson, Hugh Glass, and Jim Bridger, to name some, maybe television westerns, too. The release of the movie Jeremiah Johnson in 1972 was perfect timing. I probably watched it twenty times in just months after its release. Western romanticism claimed the day. I attended college at Southern Oregon University where I hiked and fished in the Siskiyou Mountains, after which I taught English at a little high school nestled against the Cascade Mountains where I jumped at the chance to teach Literature of the American West.

We lived near the North Umpqua River where spring and summer meant fly fishing or trekking off the highway into the towering Douglas fir with a backpack and sleeping bag to find a place with a small opening for a view through the tree limbs to the stars. There I'd sit by a campfire and watch for griz. Never saw one, but the wilderness fantasy world was alive.

When we traveled north to Alaska in the late 1970s with our baby boy, Jack, I carried with me paperbacks by western writers Zane Grey, Louis L'Amour, Don Berry and A.B. Guthrie, to name a few. About this time, I first read Edward Abbey's *Desert Solitaire*. Wilderness is not a luxury, Abby wrote, but rather a necessity. Jedediah Smith grabbed my attention, the mountain man who survived a grizzly mauling. We named our third son Jed. At work, I day-dreamed of hiking through rocky mountain passes, wild grasslands, and along tumbling clear-water creeks, but the imagery of the wilderness often clashed with locker rooms and cafeterias at the high school where I worked as a teacher.

Most people who live in Alaska have handled this question: *What brought you up here?* Recently, I told a work partner I moved to Alaska forty years before to live in the wilderness. She shook her head and turned up her lower lip in what had to be disapproval. I understood. Solitude has been generally thought of negatively. Strange, isn't it, to want to live far from people.

I studied wilderness living anyway.

Obstacles blocked the way. Foremost, I needed a place, a piece of land. I spent years thinking about this. I could sneak out to the woods, maybe along the upper reaches of some wild river and squat on a piece of land. Alaska's huge. I'd just use it for a while. I knew it was done. I worked with a teacher whose husband's job was to root out squatters for the Bureau of Land Management, but squatting didn't fit me. I could buy a remote piece of property, maybe with a cabin already there, most of the work done, but, of course, I didn't have any money. I could homestead, too, like the pioneers, stake a homestead, prove up on the land until it was mine, but I wasn't a farmer or rancher, and besides, that's not the work I wanted to do. I wanted to just live in the wilderness, gaze upon mountain peaks, rushing rivers, hunt, fish, and maybe tend a garden. The door had closed on homesteading federal land anyway.

The Federal Homestead Act, enacted in 1862, became one of the foundation pieces of the settlement of western America. More than ten percent of the land in the United States was settled under this law. From the Act:

> Be it enacted, That any person who is the head of a family, or who has arrived at the age of twenty-one years, and is a citizen of the United States, or who shall have filed his declaration of intention to become such, as required by the naturalization laws of the United States, and who has never borne arms against the United States Government or given aid and comfort to its enemies, shall, from and after the first of January, eighteen hundred and sixty-three, be entitled to enter one quarter-section or a less quantity of unappropriated public lands . . .

The program offered large quantities of free land in the west resulting in the discovery of valuable natural resources, the building of towns, and a railroad that reached California. The act was expanded to Alaska in 1898 but homesteading in Alaska proved very difficult. Fewer than two hundred homestead applications were filed in the first fifteen years after expansion to Alaska. During World War II, soldiers saw Alaska first hand, and there was a boost in applications after the war. The Federal Land Policy and Management Act of 1976 repealed the homestead law, and after it was all said and done, more than three thousand homesteads had been approved in Alaska, totaling more than three hundred and fifty thousand acres.

After the federal homestead program ended, Alaskans applied pressure at the state legislative level to increase state land offerings. In 1980, there began a significant increase in state land sales programs and more than two hundred thousand acres were sold to Alaskans.

Here was my chance.

I found my land in a mid-1980s lands brochure that grabbed me and wouldn't let loose. I read it every night. Published by the Alaska Department of Natural Resources, the document described staking areas across the state. The brochure included a map and description of the land, an estimate of the price, requirements to gain ownership, and some comment about access, such as "snow machine in the winter" or "fly-in only."

An enchanting area in the central interior described in the brochure sounded perfect to me: accessible by boat, with ten thousand acres of wilderness surrounded by millions of acres of more wilderness: Bear and moose roam this remote country; waterfowl congregate in massive flocks, and bald eagles nest in the tallest white spruce. Hawks live and hunt along the swamps, and within the shadows of the highest branches, the land is ruled by the owl, the master raptor of the forest.

I embellish this description a bit thinking back. It was the vast area south of the Yukon River and north of Denali.

I often fell asleep with the land brochure. It spread across the bed at night. Doylanne, propped up in bed reading, listened to me build cabins, all resembling castles, while three young boys slept in a room a few feet away and an infant boy dozed in a crib at the foot of our bed.

To be clear, the state homestead program was not a replica of the federal program. The state program limited parcel sizes to forty acres, and along waterbodies, only twenty acres were allowed per parcel. In order to gain ownership, the land-staker could do the following: 1. Build a cabin, survey the property, and live on the land three years within a five year period; 2. Build a cabin, survey the property, and buy the property at a reduced rate; or 3. Survey the property and buy the land outright.

I wanted the first option.

"We could just live out there," I'd say. "Great place to raise kids." I'd catch her looking at me, her dark eyes catching the light from our reading lamps. She'd be smiling, but I wasn't sure if she liked the idea or was just amused that I was talking so much. I'd go on and on, and when the lights went out, I closed my eyes and watched for moose amid lupines and flaming fireweed until I fell asleep. On the weekends, I sought out boats, attended outdoor shows and imagined the river turns.

I read all I could, enjoying Alaska stories, and admit, I feared at times some calamity. After all, I regularly read the news. On March 30, 1987, an Alaska Airlines 737-200 lifted off from Juneau and collided with a large fish. An eagle, with a fish gripped in its talons, crossed flight paths with the jet. The eagle escaped injury, but the fish smashed onto the windshield.

Sometimes, Alaska does that to us, smashes us. I thought often about recklessness, a flaw that followed me around. My life was driving by watching the hood. Who cared about what lurked down the road?

The staking area was more than five hundred miles from home, three hundred miles over cracked and rolling pavement to a boat launch, and then more than two hundred miles of river. Round trip, nearly eleven hundred miles, roughly the distance from Seattle to Los Angeles. Far away into the wilderness, and that's what kept the dream alive. Who has seen the distant layers of blue-mountains while driving and not imagined going to them?

Wilderness dreams seemed everywhere. At the high school where I worked after moving to Alaska, I hung at lunchtime with a group who talked constantly about high school sports and hunting and fishing. Any extra time we griped about the principal. I was the head football coach, so a good share of the time was spent telling me what plays to call and why the 4-3 defense wouldn't work, and I planned one day to become a principal, so the snide comments were interesting, but mostly, I was interested in the outdoor stories. If half of the stories I heard were true, I was certainly privileged to lunch with so many renown outdoorsmen, all of whom had their own wilderness dreams, and I suspect most of them, like me, had some desire to escape the general craziness of the world, to pull the door shut and allow the noise to fade. Robert Frost described it in his poem *Directive*, "Slowly closing like a dent in dough."

"We can't afford this, can we?" Doylanne asked. I sat at the kitchen table working on a list of purchases for my first trip to the staking area. The list was growing.

"No, probably not," I said. I didn't care about money.

Doylanne, rinsed dishes. "You know, you should care more about that."

"About money?"

"Yeah."

"I didn't say, I didn't care."

"No, but you were thinking it."

Had me there.

I waited for the ice to melt. The spring of 1987 was normal for the north during those years, which means it drug on and on. The snow wouldn't thaw, clinging to everything it touched. I cleared pathways to the soil across the yard to give the sun a better chance. The driveway was manicured and ready for the big drain, but the cold held on, freezing every night and warming only a short distance above that mark during the day. I awoke most nights, too attached to sleep to get up and look, but my first thought was the temperature. Was this the night it would break? It was only me, of course, anxious and impatient. This is the way it always was. Winter never wanted to let go.

In early June, just days before the first trip to the staking area, my brother Charles came to town to join me on this trip. Charles walked with an easy saunter, head tilted a bit. I loved watching him play high school football. A one hundred sixty-pound halfback, he'd lope back to the huddle like he'd just scored a touchdown, every time. He scored a bunch of them, too, but it didn't matter whether or not he scored, he just ran that way, walked that way, too.

After lots of list making and head scratching, Charles and I picked out an eighteen-foot army-green, flat-bottom boat in Anchorage. We hurried. He took time off work and came into town with two outboards, a new twenty-five horsepower two-stroke outboard and a rebuilt twenty-five two-stroke for a backup.

The gasoline powered outboard was invented by Cameron Beach Waterman when he was a student at Yale in the early 1900s and made commercially successful by Ole Evinrude just a decade later with a three horsepower two-stroke outboard that sold worldwide. A two-stroke engine completes a power cycle with two strokes of the piston (up and down) during only one crankshaft revolution. Oil is mixed with the gas as a lubricant and burns with the gasoline. What's twenty-five horsepower? In the late 18th century, engineer James Watt used the term to compare the output of steam engines with the power of draft horses. Twenty-five horses are a lot of horses, but a twenty-five horsepower motor weighs about one hundred pounds and is considered a small outboard.

Charles and I studied maps and translated miles to hours and hours to gallons of gasoline. Calculations went like this: Two hundred miles at fifteen miles an hour meant about fourteen hours of running. The outboard used about three gallons an hour, so we needed twenty-eight gallons for a one-way trip. We needed to run around some once we got there, so maybe we could get by with thirty-five gallons. We settled on seventy gallons for the trip, thirty hours of running. We guessed like ordering at a French restaurant. Two hundred miles might have been much more, or less, (after all, we came up with the distance with a string on a map) or the outboard might burn two gallons per hour rather than three. Charles led the gasoline calculation exercises. He taught business and math at Kuskokwim Community College so was more qualified than an English teacher. He said seventy gallons and six quarts of two-cycle oil. Gasoline was going for ninety cents per gallon at the time. The gasoline weighed more than four hundred pounds.

We tackled the supply list and the weight: Two adults (four hundred and twenty-five pounds, most of it me), two motors, seventy gallons of gasoline, camping gear and supplies for a week's trip, and tools to stake a homestead. At a nearby lake, we loaded the boat with containers of water to approximate our load and launched for a test run. The boat rose steadily, and when we both leaned forward, it broke over the top onto step.

When a boat barges through the water, it pushes the water ahead of it, meeting significant resistance. *On-Step* is getting the boat to ride mostly on the surface, thereby reducing the resistance and improving the boat's speed. Speed pushes the boat to the surface. Flat bottom boats, like the jon boat, with enough power and enough water can jump on-step easily. The boat rose and cruised on-step. We throttled back a bit. Not quite enough power, but it worked. The motor strained and exhaust containing carbon monoxide and nitrogen oxides dissipated behind us and unburned oil leaked into the blue waters of an Alaska lake.

We cruised across the lake a few times watching the mountains topped with the last snow of the year.

We were ready. It was time to go find this land. I held infant Jake, and the other three boys took turns hugging me and dashing around in circles in the driveway. Doylanne helped Charles and I load the pickup and boat.

"Would you be careful, please?" she asked.

"What are you going to do while I'm gone?"

A faint facial twitch, then she smiled. I knew that look, too. I'd just asked a stupid question. I handed her baby Jake.

Charles and I pulled out heading north from the Matanuska-Susitna Valley heading through some of the most beautiful country in the world. Driving the Parks Highway north is like running a river, curvy, unpredictable, lined with trees. A bear might dash across the road. Moose, often standing near the trees, might get in the way. I would, on one trip a few years ahead, count seventeen moose in three hundred miles.

Six hours later, we arrived at an empty boat launch. A good-sized river flowed before us, about three hundred yards across. Not fast, but gurgling, bubbling, churning, powerful, the color gray in the sunlight and green near the far bank in the shadows of the foliage. A t-shirt day with persistent mosquitoes, we hurried the launching like it might go away. We soon headed downstream with the current helping to propel the heavily loaded boat onto step.

It all felt good as trees, mostly white spruce, the northernmost tree species in North America, clicked by, and we tried to calculate speed. Fifteen miles per hour, maybe, maybe not, but we were on-step.

Charles ran the boat. I held a topographical map in my lap and traced the route. Probably hundreds of river bends, and maybe thousands of sandbars lay ahead, we didn't know. Before long we ran aground, surprising both of us. We ran up on the sand and were stuck in about four inches of water fifty yards from the nearest shore. I pulled on hip waders and jumped out of the boat and shoved us free. There were times during the day when we both got out and pushed.

Over the next hours, the river turned back on itself, ran all directions, but the progress was west and south toward the Alaska Range and Denali. I followed the boat's progress, my fingers hopped along the curves of the thin blue line representing a river.

My mood darkened on some stretches where shadows crossed the river, and we ran close to the bank. Some higher banks lined with sweepers, trees and brush growing out into the water, created danger for a small boat. The river would swallow a boat caught in the sweepers. Whirlpools formed and swallowed almost instantly. The river surface changed color as we altered our angles on the water. By early evening, the ambient temperature cooled as cumulonimbus clouds filled most of the sky.

We measured the trip six gallons at a time, the capacity of our gasoline can. When the motor sputtered, we rapidly poured mixed gasoline from the red plastic gas cans to the gasoline tank, always spilling some. The motor cranked incessantly along for three hours on a can of gasoline, what we figured, but it took us longer than expected, eighteen running hours to reach the staking area, not fourteen. The outboard ran okay but hot. The water from the spit hose was too hot to touch.

Outboard engines are cooled by water. A water pump with a rubber impeller, located in the submersed lower unit, forces water through a metal jacket encasing the motor. The motor pushes out into the water most of the exhaust with some unburned gasoline and oil. The boat operator can glance back and watch the water pump working and can also reach back and test the temperature of the water.

Charles often reached back and frowned.

We camped that first night on a beach the color of pine boards. I paced around the fire while we talked about the outboard. I fell to sleep listening to the crashing sound of sand caving into the river.

We left early, struggling upstream in shallow water until we arrived at the mouth of the river that ran through the staking area. We landed on a long beach on the starboard side. We had a task to complete: For months I had talked about the trip with a friend. He said he and a pilot buddy would fly up, weather permitting, from Anchorage and check on us if there was a place to land a bush plane. Charles and I flagged an uneven eight hundred feet of sand with red ribbon, marking a crude landing strip.

We finished with the air strip and went further up the river. Just as heavy rain hit, Charles nudged the boat softly to a beach within the homestead staking area. I stepped off onto the warm sand and my jaw relaxed. We didn't see another boat during the trip. The total trip from home took thirty-six hours, half of those in the boat.

We rushed to set up the tent where we hid and watched the rain poke holes in the river. I laid back on my sleeping bag, teeth aching, and thought about home. Neither Charles nor I had any way to contact anyone. I missed Doylanne and the boys already. They would be in bed by now. I remember big drops and a flapping tent.

The storm passed quickly. I studied the map at the edge of the beach. Facing south, the map showed the great Denali ahead, but I couldn't see it. The turns in the waterways made some sense, but it dawned on both of us there was no certainty this was the right river. At about 1:00 a.m., in dwindling daylight and under a sky with figurines, Charles caught an Arctic grayling on this river. He held it up for me to see. The beautiful fish flashed silver in the muted light along the river. He released the little fish.

Charles beat me out of the tent that morning and had a fire going when I stepped out to a long arching light and a crimson sky. We headed upstream after downing a pot of coffee looking for a definitive sign we were in the right place. On the shallow, clear river, we soon came to a fork. The map showed the main river, the thicker blue line, bending to the east, but the water flowed slower and the surface was brown. The branch to the southwest looked deeper and faster. But I stuck to the map. The dark line always meant main channel.

I'd learn.

The boat rode high in the water, but the prop began to scrape the bottom immediately. After the first turn, we faced a massive golden eagle standing on the beach. Charles scrambled for his camera. The eagle leapt, wings beating, and swung upriver, disappearing over the trees. We soon came to a beaver dam spanning the entire river, probably fifty feet wide. We stood on the slippery sticks, and together, with considerable effort, pulled the boat onto the pond above the sticks. We putted along further and came to another dam and again pulled the boat over the rise. Soon another dam rose two feet above the river's surface. We stopped and considered the beaver-built levee before us.

Charles pointed to a great-horned owl with its tapered tufts of gray feathers perched on a birch limb extending over the river. Its head swiveled to follow our movement in the boat. I took off my shirt and laid back and watched, and Charles snapped photographs.

"I wonder if we could just run up the damn," I asked Charles, who was like a father to me. Six years older, he checked on my grades in school and encouraged me to go to college. He tossed his head to the side and studied the slippery sticks.

We were light. A chainsaw, a couple rifles, lunch and us.

"Maybe," he said.

The owl watched while Charles, in his railroad engineer shirt, straddled the rear bench seat to get in the best position to control the tiller and maneuver the outboard. He accelerated the boat onto step straight at the beaver dam. I leaned back toward the stern. Just as the bow ran up the sticks, I shifted my weight forward while Charles jerked the outboard lower unit out of the water. The front end of the boat plopped forward, slamming above the dam, and the boat advanced with just enough space to drop the lower unit. We glided forward for a few yards and stopped. Easy. We used this technique several times during the day.

We putted slowly up the stream, and I day dreamed. Once, Charles, working for a survey crew while on break from college, stayed remote in a small trailer near the construction project. He called the house, "Eric, can you come out here for a night? You won't believe this fishing." I found a ride. "Watch this." From the bank of this brilliantly clear, small Oregon river several miles above tidewater, he cast a brown-bodied fly fifteen-feet, and I immediately saw it. A torpedo blasted from the darkest blue and shot out of the water with the fly. "Here, your turn," he said. He'd caught a blueback trout, a juvenile steelhead. We caught big fish all evening.

Charles encouraged me to experience and love the outdoors.

On the Alaskan river, we kept heading southward toward Denali. The access was much too hard, but the prospects fascinated me. We measured future effort and expense and analyzed what ifs. What if an airstrip could be built? What if the water was just unusually low this year? What if Shangri-La was just around the next bend? We encountered seven beaver dams, and Charles hurdled each one. The journey continued until the stream narrowed to the width of the boat in a vast expanse of head-high grass. We stopped when we could go no farther. We breathed wood smoke, and my red skin ached. Several years later, I found this spot again, coming from another direction while hunting.

I saw an ideal cabin spot shortly after we turned downstream. High bank and the terrain rose away from the river. Wild roses and ferns grew with white spruce and large birch across acres and acres. It was easy to imagine a frontiersman's log cabin among the trees. We stopped for lunch on a small beach across from the cabin spot. I laid shirtless against driftwood near the water away from most of the bugs, looking across the river through the leaves to azure-cyan blue and patches of ivory clouds and dreamed about a log cabin: Doylanne searched for rocks along the small river, striding barefoot. On the high banks where brush was sparse, she wove among the white spruce and stopped to pick a bouquet of wild roses. She pointed to the largest paper birch and said she wanted one of those in her yard. She peered into the woods and asked if bears ate roses. The boys swam in the river and skipped rocks across the surface.

Dreams. I got lost in the light rays filtering through the branches at the river's edge and bouncing back to space.

But we had to keep going.

Charles jumped each beaver dam on the way back to camp. The technique with the boat and outboard was similar downstream, but an easier leap. Charles popped the boat onto step and as soon as the bow extended beyond the sticks, he pulled up the lower unit and hoped for adequate clearance.

On one corner, six spruce grouse, the bird that produces the ubiquitous wild morning call of the north, pecked at the sand. With a small caliber rifle, I shot two, cleaning the birds on the beach by stepping on their wings, laying one hand on their head and pulling on their legs. The entrails pulled away, leaving breasts.

We were back at camp an hour later.

"You think Doylanne'll like this?" Charles asked. We sat in the sand.

"Don't know."

"Lots of bear sign."

I slapped at mosquitoes. A week was a long time for us to be apart. I missed her. "She's tougher than she looks." Charles was asking a good question, though. I didn't know.

A bear visited late. I stood at the edge of the beach casting a mosquito pattern fly across the river near the opposite bank. I heard a noise behind me. A black bear cub stood in the wild roses. The cub, charcoal with a tan snout, slipped back and disappeared. I inched gingerly over to the rose bushes and squinted into the forest. The little bear left a trail of pink petals from the Arctic Rose, the most spectacular flower in the boreal forest. I put a rifle near the campfire. I didn't see the cub again, but later, an adult black bear swam across the river just yards above the camp. I rushed for the rifle, but the bear climbed the bank and disappeared in the brush before I could get it in my scope. I woke up several times during the night to splashes and thought about bears.

The next morning, we left early for the other fork in the river, the lighter line. The river brightened dramatically as the elevation rose and stands of white spruce along the banks thickened. We ran into the sun and sparkles. After a couple hours dodging logs and seeking some conclusive sign, such as a ribbon leading to a survey monument, we saw a path along the bank, disguised by growth but clearly a walkway. While Charles tied the boat to a young birch at the trailhead, I grabbed a rifle and headed up the trail. A small log cabin, hidden under spruce a couple hundred feet back from the river, held together with crude saddle notches made with an axe, rose from the bushes.

We knelt and duck-walked through the four-foot tall doorway onto a dirt floor and sat for a few moments on spruce stumps. No furniture, nothing on the unpeeled log walls, no windows. The only light came from the doorway and the cracks between the logs. Just outside the door, moose hooves hung from a log beam tied between two spruce trees. Who built this, hidden under the trees, far back from the river? How fortunate to find this neglected path with an old story and a decaying cabin built with an axe.

Charles and I stood at the water's edge. We'd gone too far. Along this stretch, trees, seventy feet tall, crowded the banks, shading the river. Streaks of sunlight created distinct lines across the water. Warm and sleepy, we drifted back to the camp to save fuel, the push pole keeping us from the banks and free from mosquitoes. Ducks popped out from growth on the river's edge and frantically took off to lead us away. I'd never been so confined with ducks. The siren sound surprised me. Ducks flying in formation, rounding a river corner, cutting through the air, created the sound of an ambulance siren.

About an hour from the old cabin, Charles saw a red ribbon dangling from a birch. I crawled through the roses. Someone had been here. We hiked through the brush hoping for corner posts, and we did find one near the river. They strung ribbon back into the woods but hadn't cleared a line, and the ribbon petered out. This was a spectacular spot, but they worked in vain. To stake land, we needed our precise location on the earth. Our guide, a topographical map made from aerial photography taken in 1954, was extraordinarily vague.

We got back to camp late and buried our feet in the sand. The midday temperature exceeded ninety degrees. The country was beautiful and wild, and I was astonishingly lost. Near the water, we studied maps and enjoyed the cooling air. About midnight, the wind picked up and heavy rain fell. I read in the tent and heard Charles fall off to sleep. The silence of night along the river was not quiet. The river, breeze, and caving sand competed with birds and insects. All were noisy.

Morning brought more gasoline calculations. The little motor exceeded expectations. There was enough fuel to go back up river at least one more time and get home, probably.

But there was a diversion: Mid-morning, a small red airplane on wheels appeared about two hundred feet overhead. My friend and his pilot buddy flew over us. The plane circled a couple times before dropping to tree-top level. A small package sailed out the passenger side window landing in the river a few yards upstream from the boat. Charles grabbed the package, a small box wrapped in grocery sack paper. The plane flew to the north, and we followed full throttle. After ten minutes zipping down river, we arrived at the beach where we had flagged several hundred feet for a possible runway. No plane. We circled near the beach and I opened the package and found a pint of whiskey and a note. The message read:

"Strip doesn't look good. We're going home."

Charles lifted and shook the gas can.

We traded gas for whiskey.

We headed back up stream smiling. They found us with the coordinates from the staking map I'd given my friend, and the plane had the electronics. We were there. It's a long flight from Anchorage to the area of the homestead in a small plane, passing North America's highest mountain through a pass where pilots plan to just get kicked around. I'm happy friends tried, and glad, too, they didn't try the sandy runway that resembled a strip of fried bacon. The staking map showed a survey monument about five miles upstream from camp along the river's west bank, our best hope. Using the topographical map, turn by turn, we again traced our way up river.

On a turn, where we agreed the monument must be, we began hiking through the thick brush along the river edge.

Within minutes, I found the survey monument. The numbers matched those on the map. Ah, not lost after all. We now knew where we were. The monument, about the size of a tea saucer, was engraved with information that identified its location on earth. The Bureau of Land Management survey monument marked public land boundaries. In much of wilderness Alaska, the section and township lines are not brushed. Where we stood had been surveyed by a helicopter crew, dropping in and placing a marker, moving on two miles and placing another, and on and on. Using the township line for a boundary would cost me more money and time. A surveyor would one day tell me his work had to be more exact on a township line. I didn't care.

I cleared a narrow pathway from the boat through roses and willows and hauled the staking tools triumphantly onto the river bank. In the process, I knocked Charles' new camcorder into the river. He grabbed it, but it never worked again.

With a measuring tape and a hand-held compass, we plotted the parcel. Corner posts, squared off trees with four-inch diameters, were set with identification. We engraved my name and address on aluminum cans and nailed them to the corner posts. We brushed the property lines and strung ribbon along the perimeter. After a long, hot day with mosquitoes, I stood facing the river, chest extended, axe on my shoulder, tools piled at my feet.

Charles circled the boat a few times in front of the property and snapped a few shots before we headed back to camp. What would be next? I knew it would be a year before seeing this spot again. At the camp, we laid flies against the far bank and caught grayling until it was cool enough to read in the tent.

We struggled getting home in the shallow water. The engine began to sputter about a half mile from the launch. Out of gas. I poured white gas, fuel used for the lantern and cook stove, into the gas can and a sputtering, smoking, polluting, dying motor propelled the boat to the highway side of the river. I hauled myself up the bank and hiked through the forest to the highway and walked into town to fill a can of gas.

I could never have made the trip without Charles, and he donated a new motor and camcorder to the cause. His outboard never worked again either. I asked him years later if he remembered what happened to that little outboard. He didn't.

"Scrap pile somewhere. It was worth it," he said.

That winter I taught school, drew cabin pictures, and dreamed about wilderness living, and I worried. Ralph Waldo Emerson wrote about the pressure people feel when others try to make them be someone other than who they are. "Society everywhere is in conspiracy against the manhood of every one of its members," he wrote. I suspect most of us feel that conspiracy one time or another. Doing what society demands is not always the best choice. I was a school teacher living in town with a mortgage, car payments, wife and four kids. I wanted to be a good husband and dad, but I also wanted to live in the wilderness. Would that work?

"Doylanne, you'll love this place."

"Are there swimming holes?"

"Sure, all over the place."

"The boys will like that."

I liked those questions. I'd be going back in a year, and I'd look for swimming holes.

"Is the water warm?"

6

Hum of the earth

I would soon experience the rare mystifying feelings of aloneness in the wilderness.
 "We could live there, Doylanne. Pretty easy, I think. Clearwater river, fresh air, snow-capped mountains, beautiful beaches. The boys would love it."
 "What do we do about school?"
 "I'm a teacher."
 "You're an English teacher."

 Ah, the lowly English teacher. That was our joke. For some reason I went along with it. I'd been buried under student papers for ten years and it seemed I knew nothing else.

"What's more important than reading and writing?"

"Probably math and science."

"No, it's not." She was playing with me now. "It's not like I can't do math, anyway."

A year passed and in June a friend went along with me to visit the country and help with the logging. He ran the chainsaw, and it was clear he knew what he was doing. We cruised the property for satisfactory trees and flagged straight white spruce. A cluster of spruce stood about two hundred feet from the river, and that's where we focused. The largest trees were more than two feet in diameter at the base. The ideal log we sought would taper only two inches, so a twenty-inch diameter log might taper to eighteen inches. Many of our logs were much smaller, some as small as eight inches on one end and six on the other end. I tried to measure tree height by walking away from the tree base to the spot that I believed corresponded to a forty-five-degree imaginary line from where I stood to the tip of the tree. Then I stepped off the distance back to the base. Most of the trees were fifty to seventy feet in height.

The state record white spruce tree located near Fairbanks towered one hundred five feet. These trees in the subarctic grow in an extreme environment, and the rate of growth is remarkable. Researchers have found the Alaskan white spruce produce the same amount of wood as white spruce in other northern regions of North America but do so in half the time.

We disregarded author Don Berry's admonishment in *Trask*: "This tree wishes you no harm. Please go around." We fell, limbed and bucked all the trees needed for the cabin, about twenty-five trees. Some of the trees produced three building logs, most only two.

I tramped through the fallen trees, climbed on them and ran their lengths. A lot of dead trees. The real work was about to begin. We peeled the bark from the logs in the brush where they fell because we had neither adequate equipment to move them nor a suitable place to stack them. The logs lay scattered across brush and roses. When we approached a log, billows of mosquitoes rose and attacked. We didn't use draw knives because the sap was running, and we could peel the bark away with large flathead screwdrivers.

We worked in a mosquito infested sauna. I pawed and scratched. Alaskan mosquitos lie low and hide away when the sun is shining brightest. But in the shade or when the sky is gray, mosquitoes are ferocious. Female mosquitoes bite. Males feed on flower nectar. There is a subspecies found in the subarctic with the alluring name *Culiseta impatiens*. We were being attacked by a sub-species of impatient female mosquitoes out for blood. Could that be right? I slapped my thigh and yielded dozens of smashed mosquitoes each time.

My friend applied insect repellent by spraying heavily on his head and face. He closed his eyes and held his breath. I sprayed the repellent into a pool in my palm and covered every exposed inch in a thick film. Neither heavy spraying nor heavy coating were recommended on the label, which said, "Reapply when mosquitoes begin to be troublesome again." That was just not enough.

There were times when a surprise breeze reached into our clearing and became the best defense against mosquitoes. But not often enough. The grayness of the sky and the stifling heat forced everything to the ground, including us. We worked from morning to late evening, and mosquitoes had their way with us for three long days. My friend's arms swung wildly over his head slapping at bugs. He would occasionally stop, drop his head and shudder while wiping mosquitoes away like butter on a counter. Buzzing mosquitoes, swear words and chirping squirrels dominated the landscape.

The only true pleasant time was under fading light encircled by roses around the campfire with wood smoke invading our pant legs and filtering through our fingers, soothing bites and washing away memories. The taste of insect repellent remained, even after a cool drink.

The logging experience was hard but fun. I expected that. I'd done some work in the woods before. In the early 1970s, I worked for a logging company in the logging big leagues. I worked like a madman as a rookie high-lead choker setter. My favorite co-worker was the rigging slinger. We spent our day down a mountain side, out of sight from the landing and the tower operator who ran the cables and pulled the logs up the hill to the landing. The rigging slinger told the choker setters which logs to grab and relayed directions by signal to the tower operator using a portable radio transmitter. The signals were beeps: Beep. Beep, beep beep. Beep, beep. I don't remember what they all meant, but they were messages like *go ahead slow* or *stop*. The only beeps I cared about were *go ahead fast* which meant to me *get the hell out of the way*. I sprinted as fast as I could away from moving logs and flying debris.

The rigging slinger was one of the funniest men I ever worked with, and always a bit off color: Whenever I'd step off to pee, he said, "Zip it up Eric. You sound like a cow pissin' on a flat rock." He had hundreds of them.

Funny, straight talking guys worked in the woods, and oh man, they were all tough guys. I liked setting chokers some of the time, about as much as one could like it, I suppose. It was work like hell, be fast, be tough, and don't ever be a baby, but I'm thankful I moved on early from frayed-legged pants, suspenders, hard hats, and calk boots with untied laces.

My friend and I lingered for a day around camp preparing for the trip back to the launch. Silent mostly, we fished and read.

A few days after we were back home, he called.

"Eric, I had a great time, but I've been sick since I got home. Greasy food, I guess."

He never went back with me.

I'd made the trip to the homestead twice at this point and was struck by the heat. Sometimes temperatures reached the nineties, and it was often in the eighties. Interior Alaska is known for its hot summers unlike the mild summers of Southcentral Alaska that I was used to.

At home, after the trip, I fretted about the logs. Would they warp? I remembered how we left the logs crisscrossing each other like a box of dropped toothpicks. I had to go back. In mid-July, I talked with Doylanne about running back to the property by myself to move the logs to neat stacks like I'd seen in the log home building books.

"You won't be able to move logs by yourself."

"I can carry most of those logs"

"Carry a log. No, you can't?"

"They're spruce. Light. They once made airplanes out of that wood."

"They quit doing that." She looked up from the sink.

"Quit doing what?"

"Making planes out of wood."

"That's because wood's too heavy."

Damn, did it again. We talked a bit. No argument but she did say, "Eric, don't go alone. This is silly." Standing at the sink, silent, fiddling with the last few dishes, she was uncertain and unhappy.

"If we're going to live there one day, I have to do it by myself."

When I got to the launch, something was wrong. People milled about, some with huge treble hooks, inches between the barbs. A guy told me a poor fellow had fallen overboard upstream, and they were dragging for his body. For an hour, I watched. Boats came and went. I heard the crackle of the police radio from the trooper car parked near the river, and I thought about going home. I relaxed, though, when the taillights of the vehicles parked at the launch faded in the dust. Must be over. The launch emptied rapidly. I stood for a while at the edge of the river and absorbed the warm, building wind, light sprinkles and the sweet smell of the evening. A man sat near the edge of the water, head covered with a hood. His torso leaned against the wind. Soon I pulled away. The man waved at me, and I waved back and steered carefully into the current. I was soon on-step with a light load, eighty-five gallons of gasoline and gear for ten days in the woods.

About an hour from the launch, just as the rain was picking up, I ran the boat aground near the middle of the river. Pushing with the sixteen-foot spruce pole didn't work. I couldn't move the boat. I stepped out of the back, reducing the weight by more than two hundred pounds. The gray water swept against my hip waders at about eight miles an hour. I dug my feet in the sand and pushed. Rock the boat and push. Rock and push. The boat moved ahead a bit and stopped. During a rest, while watching a rainbow dissolve, I leaned into the transom and the boat broke free and began floating away. I followed, and the sand turned soupy beneath my feet and soon I floated behind the boat. Pulling myself to where my chest was even with the transom, I hung on and drifted toward the Bering Sea.

My hip waders were rapidly filling. No starting over. No second chance. I pulled myself into the boat, lying out of sight for a long time catching my breath, hoping nothing, not even the squirrels, saw that. I peeked over the gunwale occasionally. That was stupid and close. I got away with one.

I droned on toward the property. After about one hundred ninety miles of river, a bear swam the river and disappeared in the brush. I was in the staking area now, a few miles below the homestead looking for a beach camping spot. I kept going to get away from the bear. After a few turns more, I pulled the boat up to a large tan beach. I set the tent and campfire at the river's edge hoping to deter mosquitoes.

A moan came from across the river, downstream. It was baritone sax, and the duration of each note was several seconds. With a rifle across my lap, I watched for the source of this new noise. I sat alone with bear tracks at my feet and thought too much. I studied the bear tracks, measured them with my hands. Grizzly or black bear? I wasn't certain. I followed them to the brush. There were bear tracks on top of bear tracks across the entire beach. Animals awakened during the quiet evening, and they seemed to ignore me. A gray jay flitted about. A bald eagle floated above in the cloudless sky.

A wistful time along the river, I thought about translating dreams to reality. Work was going well, and that was all good. It would be a struggle to give that away, but more complicated were remote homestead dreams with a spouse and children. I watched the river and thought a lot about this. I wanted to live here on the river, but Doylanne's universe didn't include the blurred edge of nowhere with mosquitoes (mine didn't either, yet), but she did talk with me about the land and the concept, and she helped me plan and didn't push back, too much. She was home with four young children, one still in diapers, and I was camping on a stretch of cinnamon colored sand in the wilderness.

I discovered the source of the baritone moaning. A large porcupine with thousands of quills and a headdress with gray highlights, exuding a sorrowful, cheerless call, watched from across the river. The tough on the outside, solitary porcupine, didn't sound tough, though. Only sad. I raised my rifle, placing the crosshairs on the porcupine. Its nose looked quite large. I clicked off the safety. It moved a few inches, still in view. Small coffee-bean eyes. I reset the safety and lowered the rifle.
    I built a rifle stand with driftwood and watched animals. A rabbit hopped out to the river's edge, ears twitching with each pop from the fire, a cow moose high-stepped carefully up the river toward the boat, veering across within twenty yards of the camp before sliding into the trees, and a red fox pranced through camp without turning its head. All this in one evening. The outdoors came alive in silence.

In the tent, I catalogued the sounds made by the river. A sweeper made a rhythmic swoosh as it bounded back from being pulled under by the current only to be pulled under again. A beaver's tail splashed. I heard a clapping noise on an inland pond, a trumpeter swan smacking its wings. There was another sound I couldn't identify. It was a plop in the water followed by silence for long seconds, another plop followed by long silence. I strained to determine if the sound was growing closer. It must be another sweeper I thought. Maybe a moose. I fell asleep following that sound.

I awoke to squirrels. Two of them ranted behind me. In dense fog, I putted the short distance upstream to the felled trees. I rigged up a couple pulleys and attempted to manually pull a log. It inched along. I soon learned a couple things: I wasn't as strong as I thought I was, and I didn't have the right equipment. I inspected the moldy logs lying on the ground. They'd already turned green! I propped the logs off the ground with support at the ends and in the middle and did the best I could to prepare them for a long, cold winter. By afternoon, I gave up for the day and sat on a log in a light breeze and dreamed of the cabin. I watched the downstream view and switched to the upstream view. I shed my shirt and covered myself with insect repellent and spent the day on the edge of the river.

That evening the sun held its place near an eagle's nest. I imagined it wanting to prolong its view of the river. I thoughtfully flipped flies. Time wasn't important in midsummer with twenty hours of daylight. I caught and released grayling eight to twelve inches long. Grayling, with their sail-like dorsal fin, played everywhere. They flashed through the dark water and jumped, spinning and catching light like the blades of a fan. I spent hours at the river's edge.

* * *

Fishing became a part of every trip. It had to. Fishing pulled me into the outdoors as a kid and never let go. A lot of people feel the same pull. More than forty million Americans go fishing at least once annually. Some go nearly every day. *Why* is an interesting question. Thoreau wrote in his Journal: "Of course it cannot be merely for the pickerel they may catch; there is some adventure in it . . ." Clearly, there's more to fishing than fish.

On one of those magnificent warm Alaska days when the breeze adds a shirt, Jack and I ran up river to the confluence of a mysterious mountain stream. He was on a short break from graduate school in the Midwest. He ran the boat, lifting the outboard to avoid logs, then dropping the lower unit back in the water when there was adequate clearance.

We were on a creek named after Jed because he caught a giant grayling there once. I watched for submerged logs from the bow. Our plan was to putt up the creek as far as we could and then drift quietly down. Logs dominated this little stream, some spanning the entire width of the creek just below the surface.

Facing a submerged log, Jack cut the motor, and we pulled the boat over for enough clearance to move on. Sometimes, a log rose above the surface, but we found a gap to squeeze through. Occasionally, the prop hit a log. Jack shut the engine down and checked the prop. Only thirty feet across, moving slowly, spruce crowding the edge, this shadowy stream was full of giant grayling. The bottom was sand, and at times of low water, the creek was copper in color, but on this day the water was high and the surface mostly olive, changing color as the boat moved between shadow and sunlight. A fallen spruce blocked off the creek completely about twenty minutes upstream.

Cool breeze through the trees, sunlight through the leaves, the smell of cranberry perfume in the air, I laid back in the boat. He stood on the bow. Shadows and panes of light affected my vision. The grayling cruised near the shore, in the shadows. Most of the flies were hand-tied black bodies, caribou tails and brown hackle. He flipped the fly to the bank's edge.

There were times during the drift, all silent, when it was obvious the world was a confused mess, not on the stream, but back in town, base and petty. Why are people that way? Jack's fly whipped across the sky descending gracefully like a feather. There was the dimple of a rising fish. A fleeting moment, but under the right conditions, hazy things became clear. No stomach knots or shoulder tension. Peace from a well-placed fly and mountain air, what I dreamt Alaska to be.

The sky was heavy with late afternoon contours. Jack's shoes shuffled on the boat floor as he considered the precise moment when the grayling would dart from the depths for the fly. I watched the fly flow on its course. Jack held the rod in his right hand and his forefinger trapped the line to the rod, and the trailing line dangled in his left hand. Jack flicked his wrist and delicately raised his elbow, hooking a grayling.

Beautiful grayling, members of the Salmonidae family, are found across the northern hemisphere, but the Arctic grayling can only be found in North America. Native Arctic grayling still survive in the northern continental United States in clear rivers and lakes, but coexisting with mankind has taken its toll. In Alaska, they are plentiful in remote, wild places areas not adversely impacted by development.

When the surface was just right we could see them against the bank. They'd dart through the band of light. One would go and another follow. Those sighting were followed by a cast.

These weren't small grayling like those caught in the main river. Jack released a grayling at least sixteen inches in length. Beautiful and spectacular fish, the biggest grayling in the world are over twenty inches.

It took about an hour to float back to the mouth, and Jack caught more than twenty fish from twelve to eighteen inches in length. All were released. I ran the boat back up the creek. My turn, I caught as many, switching to a darker fly. We lost flies snagged in the brush too high to retrieve or on submerged logs. The boat tangled with the brush here and there, and we took a few lazy, peaceful sojourns on the shore.

Late in the evening, Jack turned the boat downstream. At the confluence of the creek and the main river was a beach.

I asked Jack to stop so we could stretch our legs and take in the last sunlight. I didn't want to go to the cabin yet. We walked the beach identifying tracks and counting owl pellets. Wolves had recently trotted down this beach.

Jack then flew down the river, faster meant less wake. The corners came fast, so he stood to see. This part of the river was mostly free of logs, but sand bars lie waiting. He ran along the high bank and crossed sharply at the corners.

The prop never touched the river bottom. Jack stopped the boat momentarily, the air temperature warming us immediately. He pulled on a jacket. Full throttle again. Down river, an osprey dropped like a spear toward the water, leveled and lifted off disappearing over the trees.

Grayling aren't the only fish we seek. There is a fishing hole near the cabin where the great white shark of interior Alaska, northern pike, savagely cruise the waters. Upstream a few turns from the cabin, a mountain creek flows slough-like into the main river blending a copper hue with a steely blue, forcing the mixture against the bank. Light rods with spinning reels with six-pound test and a twelve-inch steel leader comprise the gear. We use a swivel and a bright lure with a single barbless hook.

Jake, Jed and I fished at this spot many times. One day, I dropped Jed off on the bank then anchored the boat several yards away in the slow current of the creek. He was eight-years-old. He made the first cast with a small red and white spoon. After three spins of the reel, a pike attacked the lure. Up the creek, down the creek the head broke the surface and thrashed wickedly. "Rod tip up," I advised, but Jed knew that. He chased the fish upstream in rose bushes and back.

"Can I keep this one Dad?" Jed asked. I nodded.

"Get it in first," I said.

Jed backed up the bank until the pike, a nice one with a long, slender, olive-green body and yellowish-white spots covering its sides, flopped in the sand. The mouth, duck-bill shaped and lined with razor sharp shark teeth, snapped at him. Sometimes a finger got hit by one of the teeth during the short, brutal fishing experience. Jed took his pocket knife and ran the blade through the brain.

"They like this lure Dad," Jed said, holding the rod so the red spoon dangled and bounced in front of him.

"Pretty shiny. I guess they like that," I said.

Jed thought some, "Surprise for them."

Yes, it must have been a surprise. A fishing lure is a lure, an enticement, a trap. Lures are meant to get attention and the most successful will imitate something natural and edible in the water (I'm not an expert on fishing lures but I have used lots of them) or make the fish mad enough to strike back. Artificial fishing lures have been used for thousands of years and today lures are big business. A single lure similar to what Jed used will cost a few dollars.

We moved on up the creek a couple turns, the water almost black in the shadows. Up on the bank, there was a patch of wild raspberries, so we picked a cup. I carried a rifle. Beyond the berries was a hay field. Not a real hay field, just miles of high grass. There were moose out there, there had to be, and bears too. Stunning and wild, but the visibility was poor, and it felt a bit spooky in the chin high grass. The wind was up so no mosquitoes. The mountains to the south were visible, Denali, crystal clear. There was a need for a tree stand to search this wild hay field. Another day. Heading out, we stopped at the mouth of the creek to cast again.

Jake, a little boy, who would one day, shortly after finishing college, run to the homestead alone in a remarkable trip of low water, breakdowns, bear encounters and rescues, was standing on the center bench seat ready to cast a sure thing, a floating lure about two inches long resembling a stuffed mouse. Just six-years-old, he let it go and the steel leader pulled the mouse down several inches

89

below the surface. Almost immediately a pike devoured the fly. Jake fought the pike several minutes before finally bringing it alongside the boat. The pike, nearly three feet long, clamped firmly down on the fly. I pried the fly out of the pike's mouth, slicing my index finger. Blood dripped into the river, blending and disappearing. I watched it closely, and for a few moments lost in thought, allowed the blood to drip off the tip of my finger.

"Look here, Dad." Jake watched me. He held out his chubby hands marked with razor cuts. He had given his blood to the river too. I pulled a handkerchief from my back pocket and handed it to Jake. He dipped it into the water, wiped his hands, and rung the cloth into the river.

Water, blood, rain, debris, and most everything else blends to a river. I looked up to the clouds. Rain water has no boundaries. NASA can speculate with some assurance where rainwater originates. Rain hitting our little stream where we fished may have come from the Bering Sea, Gulf of Alaska, the Pacific, or the Atlantic or Gulf of Mexico following a tropical storm.

* * *

I was lonely on the beach. I crawled into the tent. The swans were boisterous inland. They honked and slapped wings, performing a symphony overpowering the sounds from the river. They harmonized and developed a cadence with predictable distance between tones. I laid my arm across the stock of the rifle and fell asleep.

In the morning, while bacon sizzled in the frying pan, I knelt in the sand tossing small pieces of driftwood on the campfire as the smoke disappeared in the blue. The driftwood, white, weathered and washed out by rain and wind caught my attention. It dawned on me bleach would clean the mold from the logs.

I went back to the logs. No bleach, but I selected the cap logs and the ridge pole, lifted them above the ground and supported them, hoping they wouldn't warp. I cleared brush and picked possible spots for an outhouse, wood shed, and storage shed. I spent hours sitting among the logs, drawing cabin pictures, adding linear feet, devising a plan.

In the evenings, I wandered along the river's edge, sat by the fire, and listened to the ubiquitous low frequency hum of the earth. A steady breeze blew off the river and within a week the temperature dropped to the fifties. I was alone and aware of the dissonance. I was where I wanted to be but not with whom I wanted to be. The bugs hid in the brush. Behind me a ridge blended to dense black spruce and muskeg and fewer deciduous trees. I meandered in the vast boreal forest, the taiga, a nearly continuous range of coniferous trees across the northern world, a band of mostly wilderness accounting for nearly thirty percent of the world's forest.

I putted again to the cabin site the next morning, stood where the door would be and imagined it was all done, every log and detail in place. The cabin resembled the hundreds I had admired in books and movies, with logs the color of rawhide and a lantern illuminating an indoor out-of-focus feature. I stared through a window and saw a book shelf, a mirror and a kerosene lantern. Then I stood indoors watching a dancing fire behind the glass door of a small woodstove, a small stack of kindling, and poker against a wall. I heard the joyful screams of children. I opened the door, pulled chairs close, and sat with Doylanne while they built a stick fort.

I ran home without a problem, dreaming of a log cabin.

7

Look before you leap

One day a teacher stopped by my office. "Eric, got a couple minutes? Sorry about the fire drill today, but I have a funny one." I waved him in. "When the fire alarm went off, the class lined up at the door like it's supposed to. All but you know who. He was fumbling with the padlock on the bicycle chain he used to chain himself to his chair. There was no way we were going to make it out on time, so I told him to drag it."

"I heard your class was late." I was the assistant principal in charge of fire drills.

"Yeah, we finally made it outside and the custodian used some bolt cutters to get him out of the chains. This is what's funny: I started tearing into him a little and he said, 'Sorry, but you should tell us. If I'd known there was going to be an alarm, I wouldn't uh done that.'"

I remember another student a couple months later when the snow was flying. It was an open gym during lunch hour, too cold to go outside, so the kids came to the gym after eating. My job was to keep these middle school kids in order. This young guy was talking with me about his dad's tree stand where they hunted black bear with a bow. This kid had lots of hunting experience, and I was listening. We stood in the middle of the basketball court. Against the wall, beneath the baskets, were large pads to protect kids from crashing into the brick wall. This day as we were talking, without any notice, he took off sprinting toward the wall and crashed in to the pads and bounced back to the floor ending up prone on this back gasping for air. I ran over to him.

"Tyler, what was that?"

He looked up at me, panting. He didn't know.

I didn't see any blood on Tyler. "Stay put awhile. You'll be okay. When you can breathe, walk it off."

I, too, felt it was time to run full speed ahead. There were big decisions to be made. How big should this cabin be? What kind of notch? I studied the books.

I again waited for the snow to melt. That winter, 1989, had been one of the coldest people could remember. January had a stretch that squeezed much of the state for weeks. Coastal Homer hit twenty-four below. Fairbanks was miserable and dangerous. Schools closed and were used for shelters in the Matanuska-Susitna Valley. Low temperature records were broken across the state. The state low temperature record of eighty below at Prospect Creek on the Dalton Highway held, but the state low that winter, recorded in the region of the homestead, dipped to seventy-six below. So, I watched and complained until the ice and snow disappeared.

I arranged for a guy with a big boat and knowledge of the rivers to haul the building materials and tools to the cabin site. In mid-June, Charles and I hauled the building supplies to the boat launch and unloaded on the beach. We waited into a warm, dusty afternoon. I sat for hours on a box of nails reading one of Cormac McCarthy's great novels. The freight-hauler showed up in late afternoon. He said he couldn't do it, things had come up, didn't have the time.

The supplies were stacked in front of us. I trudged up the beach. An image inspired by McCarthy's book came to me:

A horse stood before me in horrendous pain, bloated, its head swollen, eyes protruding, pus oozing from its nostrils.

My head throbbed, and I pushed fingers into the base of my skull, and then . . . and then . . . as happens at times along the river, a wave of cool air swept over me, and then another, and another, and a muddy idea, one I'd considered during nights through the winter, gradually focused, revealing a plan that just might work, and I felt much better. This was not so bad. All would be okay.

I walked back to Charles.

"So?"

"We can do it ourselves." I grabbed my book off the stack of boards and thanked the author for pointing out it could be worse.

A small crowd had gathered at the launch by the time Charles steered the overloaded boat with four inches of clearance at the gunwale away from the launch and headed downstream. The thirty horsepower propelled the load, but barely, and I trailed my fingers in the river. We headed toward purple mountains and a flushed sky. Late in the evening, about sixty miles from the launch, we chose an island sandbar with pearl colored driftwood and high cut banks for a camp.

After unloading, Charles drew the short twig and headed back to the launch for another load. We had about a third of the supplies. I set up a tent and slept. By mid-morning, Charles was back with a full boat. We could do it with one more load. After moving the materials off the island into a thicket on the north shore mainland, I ran the boat back to the launch.

Charles stayed on the island to stay clear of most of the mosquitoes.

I made it to the launch late in the day and loaded the truck as fast as I could. As I backed the truck down to the boat, the truck died and wouldn't start. I stomped around the truck adding special creative layers of profanity to my vocabulary. I yelled the words across the river and an echo replied, but it soon became evident that it was all meaningless. No one was listening, and the river didn't care.

I loaded the supplies in the boat, then trekked into town and found a mechanic. Back at the truck, we both crawled under and kicked and pulled on greasy parts and agreed the starter wasn't working anymore. There was also an oil leak but that wasn't the immediate problem. He towed the truck back to his shop where I could pick it up the next time in town.

I pushed away from the oil stained sand and headed down river. Heavy and slow, I eventually reached Charles late in the evening. We moved the camp to the mainland and made dinner under a troubled sky. No darkness, there was no need to wait for morning to shuttle closer to the cabin site. Charles waved from his new camp, and I tried to get the boat on-step, but no. I ran with the current for about twenty miles and turned upstream. The pace slowed to just a few miles an hour. I stood, sat, stood, watching the shoreline all night, convinced I could crawl faster. Late in the afternoon, still shy of the staking area, I stopped. I couldn't go any further. I needed fuel to get back to Charles.

I stood on a beach on a beautiful river bend with a partly shrouded Denali view. Through swarms of mosquitoes, I hauled all the materials into the woods and dumped it on some devil's club. After I finished, I cooled down at the edge of the water, facing a light breeze alone on the 64th parallel.

I would one winter evening after this trip trace that latitude around the world and dream of circling the globe. Fewer than two percent of the world's population live on the 64th parallel and northward. Walking east from this spot, I would pass a bit south of Fairbanks then cross Canada's Yukon Territories, Northwest Territories and Nunavut. After finding a worthy boat, I would sail the Atlantic Ocean to Greenland's Nuuk, then pass south of Iceland's Reykjavik to the Norwegian Sea. Then on land again across Norway near Oslo, across Sweden, then boat the Baltic Sea to Helsinki, then overland across vast northern Russia to the White Sea and Onega Bay, the Sea of Okhotsk to the Kamchatka Peninsula, the Bering Sea and a little southwest of Nome. Circle the world.

I was in a lonely place. A raven flew away from me down river. In a hundred square miles there may have been a few people, maybe not, but if people were out there, they were likely back in the trees out of sight because in my travels in the area I wasn't seeing anyone. In the 1880s, Lieutenant Henry T. Allen, Second United States Cavalry, and his men, explored this region. He wrote about his adventures and many hardships in *An Expedition to the Copper, Tanana, and Koyukuk Rivers in 1885*. They began at the mouth of the Copper River in the Gulf of Alaska and traveled north deep into the interior north of the Yukon River, an incredible expedition. They floated, jostled and poled on rivers, portaged mountain passes, fought mosquitoes and hunger, and completed one of the most daring journeys in

American history. During the more than one hundred thirty years since Allen's exploration, a handful of small towns have sprouted along scarce roads in the vast region. A few thousand people have made the beautiful, but still very remote area, home, and on the region's eastern side, an oil pipeline runs through it. In the summer of 1906, explorer Charles Sheldon, traveled along these rivers too. He wrote about a trip into the area in The Wilderness of Denali:

> . . . a silt-laden river with very muddy water two or three hundred yards or more in width, meandering with rather sluggish current through the broad, flat, swampy territory . . .the weather very hot, and mosquitoes were swarming, and the horses, cramped in a narrow space near the boiler, were continually tortured by both mosquitoes and big horse flies.

Indeed.

I took a bath in forty-five-degree water. After a long count, I dove in the river and exploded out and shook like a dog. Standing on wolf tracks, I put on clean clothes and changed from hip waders to tennis shoes. It took some time before the tingling at the top of my skull subsided. I fastened my life vest, and with an empty boat flew with the current. I stopped and ate a can of sardines in the sun while the boat floated like a birthday balloon past some bluffs and cliff swallows. It took five hours to arrive back at camp.

Charles pointed out that the old camp, the one on the island, moved from just the day before, was now under water. That startled me like a near miss on the highway. I didn't know that about the river. The changes can be dramatic and rapid. I have never forgotten.

I crawled in the tent to sleep, and Charles ran back to the boat launch to get food and fuel. After a short nap, I climbed a couple hundred feet up the ridge behind the camp and waited. With binoculars, I glassed miles of river. A large open boat with a heavy load passed going along the far bank, the rooster tail from the outboard rose higher than the boat. It was the first boat I'd seen on the river in three years. Whipped cream clouds built all afternoon.

Fireweed bloomed on a hillside up river. Dark green spruce needles alternated with birch leaves that faded in places to amber. The gusting wind lifted sand from the exposed beaches in the braided river, resembling rising steam. A large furry butterfly, mostly brown with some yellow and black, landed on my knee and stayed. Not a flutter. I flexed my leg and I blew softly, but it held steady. We sat together for minutes until it spread its wings and two eyes appeared. It lifted off and stopped on a cranberry bush.

I saw Charles thirty minutes before he would reach the camp, so I headed down the hillside to meet him. We immediately left for the property. Again, we were too heavy. It rained hard, and we plowed all night through rushing water and sinking clouds. Wind and rain slapped us, and we hunched over and hid our faces from the cold. About three in the morning, we stopped on a beach and started a fire with a soda pop can of gasoline and rain-soaked driftwood. Charles cut the top off the empty aluminum can and filled it half full. He sat the can under driftwood piled in a small mound and lit it, creating a torch. The flame channeled up the walls of the can. The wet wood smoldered, eventually flamed, but the rain pounded us. We huddled over a struggling, pitiful little fire without a chance to burn. It felt like it might snow. We talked about moving forever from Alaska. California here we come. We stomped around the fire and swore. Long moments, though, were spent in silence watching the helpless flames. The rainwater rolled off sleeves and onto our hands. Drops gathered at the tip of the nose before the current forced them off the precipice. It was maddening and cold. After shivering around the fire for more than an hour, we trudged back to the boat and kept going.

On the morning of the fourth day, after running all night, with the sun peeking through rapidly breaking clouds, we set up camp about ten minutes below the homestead property. We erected a wall tent near the edge of the river, the opening facing upriver and the breeze. We set out our sleeping bags and slept.

I awoke first and climbed through thick willow to mixed deciduous and coniferous forest to the summit of a ridge. There I could see some of the country. Trees hid the river, but I could see that the drainage meandered north through rolling hills for as far as could see and to the south to yellow savanna and mountains. I sat against a birch and imagined hiking to the multi-colored mountains to the south. There would be a lake to circle, a meadow to traverse, a stream to follow, miles of brushy hillsides to maneuver before reaching tree line and a ridge leading to the snow. Then the way would be mostly vertical to the summit and the sky with whipped cream clouds.

I hiked back to camp and prepared to leave for the materials and fuel cached one hundred twenty-five miles away. Charles had built a huge, roaring driftwood fire. I watched the flames for a while thinking about Doylanne and the boys. They would be wondering.

I left with a light load, feeling good running full speed into the pink sky. The propeller touched nothing but water. I reached the materials in six hours and within an hour headed back to Charles with a full load.

Slow again. I slogged upstream and counted the corners and sweepers to stay awake. I saw a cow moose and two small calves on a beach just above a swimming pool clear stream.

The mother nudged and rubbed against the Irish Setter red calves. They seemed to be all legs. They didn't slide in fear into the brush but watched me plow ahead. I vibrated on, and after ten hours against the current the smell of a hot motor forced me to shore. The motor was pumping water too hot to touch. I let the motor sit. I walked into the forest and found a burl the size of a kitchen table growing on a spruce twelve inches in diameter. A beautiful anomaly, a burl is a growth caused by stress. Something bad resulted in something beautiful. Nothing felt beautiful. There was still a long way to go. I took my time in the woods near the burl slapping at mosquitoes and watching a scolding red squirrel dash among trees to give the motor cooling time.

But I had to go. I fired up and headed up river. An hour later the water pump quit. I tossed out the anchor, laid back on the bench seat, and shook my head at a cranberry patch on the near bank spreading like fire up the slight incline away from the river. The rose was holding strong, petals still intact.

This outboard motor let me down. Piece of shit. Bolts, nuts, mysterious parts, pivoting over mountings on the transom to control direction of thrust. That's what motors do. Eventually, always. I didn't have a spare pump kit. This was, indeed, a problem. I studied the berry patch and the bank leading into the spruce. It looked like a good place to leave the materials.

We don't have much control over what happens to us. When working as a choker setter, an old hand, who was probably in his late thirties gave me some work advice. He told me, "Here you do something, even if it's wrong." Dubious advice, perhaps, but it stuck with me. Don't overthink it; get off your ass and do it, and stuff will happen.

That's what I got from it. Sometimes, of course, stuff happens anyway. Maya Angelou said, "You may not control all the events that happen to you, but you can decide not to be reduced by them."

I heard another outboard coming from downstream. I couldn't believe it. They were painfully slow, but a canoe with two men came into view. They pulled beside the boat. Bear hides covered the front of the canoe. The operator wore a beaver hat and aviator goggles. The outboard, a vintage Johnson seven horse, powered a weathered square back canoe. We talked about the river and bears. They were heading further into the mountains than my place, at least two more river changes, two or three more days at their pace. Nice guys who seemed genuinely concerned about me, but there was nothing for them to do. We waved goodbye, and they struggled to make progress against the current. It took them thirty minutes to round the first corner upstream, and I heard the whirr of their outboard for nearly an hour. I poled the boat to the cranberries and roses and carried all the building materials into the trees.

I carried a twenty-five-horsepower outboard reconfigured by Charles' son, Mike, from salvaged parts given to him by neighbors trying to get rid of their outboard parts piles. Pulling and pulling on the starter rope yielded nothing. It sputtered some. Sometimes billows of smoke escaped through the exhaust. Late in the evening, I kicked the boat free from the beach and began drifting as I worked on the engine. Not many options. The engine would function, or I'd get rescued, and there was a greater likelihood of encountering river traffic near the mouth, seventy miles away.

I used the pole to stay toward the middle of the river and away from the sweepers and mosquitoes. I pulled the starter rope hundreds of times and the starter assembly broke. I took the top apart and the pull rope spring uncoiled in my lap. I kept going. I wound the rope around the crank head and pulled. All night the boat drifted, and I never stopped pulling. Sometimes, the motor chugged, spit, and belched.

I watched a bear study me as I drifted along a stretch of high bank and spruce. The black bear didn't run away. It stood between a couple spruce and turned its body to follow me down the river, bounding softly through low bush cranberry plants and roses. I didn't grab for my rifle.

* * *

Nearly every trip through the years, bears and berries occupied part of our time. When the low-bush cranberries became ripe in late August, we often ran down river a few turns from the cabin and hiked to a small horseshoe-shaped lake where fingernail-size berries grew with the moss and blanketed the ground. Once, about 1999, our black toy poodle, Princess, went with us in search of berries. From the takeoff spot along the river, an eastward compass bearing led to a lake where cranberries rimmed the edge. I worried about bears that day. Maybe a dream was the source of the worry, or maybe the feeling had a more practical basis, such as we were in a berry patch in bear country. We picked berries as we made our way. Princess followed us on a leash. The poodle, seven inches tall at her front shoulders, couldn't keep up so the progress was slow as we meandered through the forest. When her leash caught on a branch, she'd lie down. If she ran into a clump of thick grass, she stopped and waited to be rescued. At the berry patch, we spread out and picked in small cups and poured the berries into a plastic gallon container. Our heads swung up at snaps in the woods. Red berry juice stained our fingers and clothing.

A pair of swans with four cygnets paddled into view. I'd never seen four before. I edged closer to the lake with a camera and rifle. The poodle followed me on the leash. The boys watched while Doylanne kept picking.

The day was hushed and uncomfortable. We had just hiked an animal trail covered with bear scat. I cleaned my rifle scope. The poodle scratched her back in the shrubbery and stared off into space. I moved slowly closer to the lake as the swans moved away. I watched for a bear too. Perhaps confirmation bias was at work. Bears like berries. We were standing among thousands of berries.

I knew when it was time to head home to the cabin. I urged Doylanne to hurry as she picked along the way. Heading back to the boat, I ran the leash through my belt loop. Doylanne walked near me, her arms full of cranberries, stooping often to top off a cup.

"There!" It was Jed. We followed his pointing finger to a black bear running away from us about fifty yards ahead. I picked up four-pound Princess, and we took off for the boat.

* * *

I drifted all night using the push pole to steer clear of the sweepers. Through the fog in early morning, I floated into the river mouth. I needed to get over to the main channel of the oncoming river where any sensible boat traffic would be, a half mile away. The river, one of the largest rivers in Alaska, scared me. As the boat entered the confluence, I pulled with everything and the engine started.

It seemed impossible.

I gunned it. Hectic minutes, full throttle, I ran the boat across the river, but as soon as the boat hit the main channel and I turned upstream, the motor sputtered dead. Using the pole frantically, I forced the boat against the shore and jumped off with the anchor.

I stood on the bank catching my breath, astonished.

I needed rest.

Look before you leap. I climbed into the boat and discovered I'd left the camp and Charles without a tent or sleeping bag. I sat on the bench seat and held my head. I felt the pulse at my temples.

A blue poly tarp eased my mind. When we left Wasilla days before, I'd stuffed it beneath a bench seat rolled and fastened with a bungee cord. Everyone in the north has one of these. I found a spot in the shade, pulled the tarp over me, and tucked in the edges to block the mosquitoes, and slept for several hours. I awoke drenched in sweat and thirsty. I cut the tops off two aluminum cans, filled one with silty river water, and strained the water through my t-shirt into the other can. The color of the water improved, and it was cold.

Back to the motor. I wrapped the starter rope on the crank head and pulled, rewrapped and pulled, and on and on until the engine started and ran full throttle for about ten minutes and died again. I was getting closer to the launch. I tightened the head bolts on the engine some more. The engine ran for about thirty minutes. I tightened the head bolts a little more. The engine started and ran, cranking out a rhythm, smoke and smell only a two-stroke outboard can do. Like Steinbeck's Tom Joad, I listened intently for any variation in the engine's rhythm. Buckminster Fuller called it "growing intimate with my machine." What's the explanation? Likely, the head gasket wasn't sealed on the outboard; tightening the head bolts solved the problem.

I made it to the launch and found the mechanic. No starter yet. I walked to the only tavern in town to find a ride. The tavern floor sloped one way, the bar the other. A couple guys played pool. Engulfed in cigarette smoke, I downed two cans of beer.

"Anyone run me to a boat shop?" I asked loud enough to be heard across the room.

"I can," he said, extending his hand, a big guy with a full head of bushy blond hair. "I'm at your service." We settled on fifty bucks and a tank of gas.

Sixty miles away I bought two water pump repair kits, a prop, a head gasket, and a case of outboard motor oil. We stopped at a liquor store, and I bought both of us some beer. On the ride back to the launch, he told me his wife left him during the winter to go back to the Midwest, took their two daughters with her. A tough time, he told me, but now he had a girlfriend.

"I'm going to stop by and get her. She can ride with us back to your boat. Okay?" I took a long minute to consider the idea. What are minutes now?

"Great. She's always ready to go. It's on the way." He disappeared into a cottage not far from the boat launch. Minutes later his girlfriend climbed into the truck from the driver's side and slid up close against me, barefoot in shorts and tank top. Pretty girl.

"Let's go play pool," she said. "Wanna go?" She asked me.

I watched the sheen of the river. "You two are doing good, real good, but take me to the launch. I need to change out the water pump."

They encouraged me for a few minutes, wished me luck, and said I had a place to stay if I needed it. After hugs, they drove away to the tavern.

I installed the water pump kit in summer morning light, first time tearing into a lower unit. People don't sleep here. Several local guys stopped to offer suggestions. One took me to a gas station where I filled a drum with fuel and picked up some food.

I called Doylanne from a pay phone in the station. We whispered like we were at the movies.

"You're not there yet?"

"Just lots of problems."

"Maybe this is a bad idea?" We both waited. "This isn't easy is it?" she asked.

"Are the boys okay?"

"They're fine but miss you. We went to the pool today. Jake has another tooth, on the bottom."

"I have to go get Charles."

"All my friends are praying for you and Charles."

We talked for a few minutes, but I rushed it. I still remember.

"I love you," I said. My ride was waiting.

"I love you too . . . Use any English yet?"

No English, but I learned two valuable skills from the helpful friend who ran me to the store: He showed me how to carry an outboard on my shoulder, and he showed me how one guy can load a full fuel drum into a boat, both important skills to master along the river. A drum of gasoline weighs about three hundred pounds.

At around 6:00 a.m., I left for the homestead. The two-stroke was running well again, oily, smoky, loud, like all motors of the time, a polluting monster.

In *Polluting for Pleasure*, Andre Mele wrote that the average boat is a seventeen-foot skiff with a two-stroke motor with sixty-eight horsepower. During a typical day on the water it uses about twenty gallons of gasoline and three and one-half pints of oil mixed with the gasoline. Some of it is burned during motor operation and some of it passes directly into the environment. Mele estimates oil and hydrocarbon pollution caused by pleasure boating totals four hundred and twenty million gallons per year, equivalent to forty Exxon Valdez disasters every year. Mele published his book in 1993 (just a few years after this freight-hauling trip on the river).

Not so good.

I made excellent time on the first miles of the trip but slowed notably after picking up the supplies along the river, too much weight for the small outboard. Moving the building materials from the brush to the boat was easy for me, though. Plywood slung onto my back, held in place with one hand, the other holding a box of steel spikes, no problem. I was near my physical peak that summer, not in aerobic conditioning but in natural strength. Old man strength. I've always been strong. Not prize winning strong or anything, just strong for my age. I had that short stocky body, more round than lean, a slightly plump mesomorph. When is a human the strongest? Strength peaks at about twenty-five then plateaus until about forty, and then falls steadily. By sixty-five we've lost twenty-five percent of our peak strength.

Above the clear water creek, I saw a cow moose with one calf. The frazzled cow didn't fawn over the calf but rather pushed it forcefully into the brush. I took a short stop at the scenic bend on the upper river to check on building materials lying with the devil's club before the final push to the campsite. I ran and ran on the river. Early in the morning, the boat often passed through a patch of silver fog caused by cool air and warmer water. The fog patch loomed ahead for some distance, so I'd bundle up in anticipation of a blast of cold air. The cold air hit, blurring my vision. The first time I saw the fog approaching, I dreaded it because I was already cold, but I was surprised. The air on the other side covered me like warm bath water.

At about 6:00 a.m., after a twenty-four-hour run, I arrived at camp. Charles had a bonfire and coffee ready. He heard the outboard an hour before I arrived. We shook hands, hugged and shuffled around the campfire telling stories. He had no idea if I'd ever return. I crawled into the canvas tent and crashed, and Charles made two runs for the materials stored in the devil's club. Eight hours later, Charles had the materials at the camp except those still one hundred and twenty-five miles away that would wait for another year.

After eight days of adventures, and fifteen hundred miles of river, we'd stacked in front of us the tools and materials to build a cabin. It wasn't easy, but it certainly could have been worse. I was a step closer to living in the wilderness.

We are fortunate when experiences change us for the better. I felt those eight days may had been good for me. I made a paper boat, a Moses basket, and followed it down the beach and thought about Doylanne and the boys. Why do we do the things we do? Among many beautiful passages in Constance Helmericks' *Down the Wild River North*, she writes: "Oh, to have a wilderness all to oneself with plenty of room--what a gift, what a wealth the healing stuff of nature."

8

Angry Squirrels

> Every creature is better alive than dead, men and moose and pine trees, and he who understands it right will rather preserve its life than destroy it. —Thoreau, *Chesuncook*

I would soon think I was dying. We rested a few hours then faced logs scattered like pencils. I mixed gas and two-cycle oil for the chainsaw, poured in chain lubricant oil and tightened the chain and fired it up. The blue exhaust ran away with the breeze. About twenty-five trees, resulting in about sixty logs, were killed the year before. We calculated linear feet and agreed there was no room to mess up with the chainsaw. We set out to build our first log structure.

Log cabin construction is all about joining logs at the corner. We'd build a no-notch style V-Plank cabin. I learned about the style from Tom Walker's *Building the Alaska Log Home*. Here's how we built our cabin: The foundation consisted of nine creosote treated posts sunk into the ground. Three logs, with notches cut, rested on those posts. On those logs, we nailed two by six floor joists at sixteen-inch centers. We covered the floor joists with plywood. We started with the larger logs and built the walls to a height of six feet.

We spiked each log in two places to the log below. We accomplished this by drilling about half way through the top log and driving a twelve-inch steel spike into the log below, cinching it tight by pounding the spike head with a steel rod and a sledge hammer.

We powered an electric drill with small generator and drilled over one hundred thirty holes to a depth of six inches. We framed the doorway and windows with two by six boards. The main window in the cabin faced the river and was a double-pane, egress window. For the roof, we used twenty-four-foot cap logs. Those provided a two-foot overhang at the rear of the cabin and extended out six feet in the front. The overhang in the front provided the framework to cover the porch. We sat the ridge pole in place, resting it on the gable ends. We nailed two by six rafters to the ridge pole and cap logs at twenty-four-inch centers, and nailed plywood over those. We laid tar paper over the plywood. Charles and I built the front door with tongue and groove two by sixes and two by four cross pieces.

We made most of it up as we went along. The only log structure I'd built before this was a toy log set. I flattened the first log with the chainsaw by running the chain vertically over the log at high speed. I spiked the log to the platform. Each log, alternating butt to tip, was rolled onto the log below and shaped for a snug fit. This always required rolling the log into place several times, sawing here, chopping there, until the logs fit together. When a log was finished, both ends were nailed to the V-Plank with twenty penny nails.

We measured progress by sections: Day one, the foundation was laid; day two, the floor finished; day three, half of the log walls raised; day four, the other half, to a height of six feet and the log gable ends; day five, the ridge pole, rafters and roof; and day six, installation of a window and construction and installation of the door. We kept on late and quit when the owls begin to hoot, and the squirrels disappeared.

We worked like children playing and the cabin shot up rapidly. This wasn't child's play, though, we used a chainsaw. My midsize saw, emitting pollutants equal to an automobile, could cut anything. The first chainsaws, developed in Germany in the 19th century, were a surgeon's tool for sawing bones. Chainsaws are scary machines. In 2012, according to the Occupational Safety and Health Administration, two hundred and forty-three workers died while engaging in tree-trimming and clearing activities. Data from the National Electronic Injury Surveillance System for the years 2009–2013, show there were more than one hundred and fifteen thousand visits to the emergency room following chainsaw related accidents. Almost all the patients were male, white, and badly cut. The most common body sites for injury were hands, fingers and the knees. Most injuries were at home.

We built the cabin in a week. As the cabin rose and became a log structure, I frequently walked around inspecting, evaluating, and admiring. Not another in the universe like this, with its logs varying from twenty-inches to six-inches in diameter, and its thousand other points of distinctiveness. It stood unique in the center of Alaska.

One late hot afternoon, while near the top rung of the extension ladder, running the chainsaw on a top wall log, my body quit. I hit the kill switch on the chainsaw as I fell hard to the cabin floor. I sucked hard for air as my torso spasmed. A vise crushed my chest.

Charles immediately knelt over me, telling me to lie still.

As I lay gasping in the sawdust, I thought I was dying.

I rolled to my back and pulled in air and the pain evaporated slowly like a water stain on a shirt. Spears of light sliced through the spruce branches reaching over the cabin. I didn't see my life flash before my eyes, but I saw instead a branch sway under the weight of a squirrel: reddish body, white belly, and bushy orangish tail, ten inches long maybe. I heard babble among squirrel friends.

I carefully worked myself to my feet, walking my fingers up the log wall. I picked the sawdust from my mouth and ran my hands through my hair, knocking loose wood chips and a robin feather. I stood on the cabin floor hurt, eyes down, trembling, embarrassed. Squirrels were noisy. I knew they were angry with me. I'd killed their trees.

Charles suggested I try walking.

I stepped off the cabin deck and wandered like a drunk to the river, each step a jab in my side. I grabbed a rifle and hesitantly inched down the animal trail inside the trees along the river. Charles trailed me. I propped myself against a tree here and there and deeply took in the spruce smell. I felt pathetic. Waste of time. Doylanne and the boys will never get up here. Some crazy juvenile fantasy. Freud was right. Charles told me to keep walking.

Hours later, with the sun finally below the trees, I cast a mosquito fly toward the opposite bank. Pain shot through me like a dreadful memory and lingered. I tried again but couldn't take the pain. Grayling frolicked, and I couldn't reach them. My soreness lasted for weeks.

We finished the cabin a few days later. We used all the dimensional lumber, except for a couple two by sixes and a half sheet of half-inch plywood. There was also a handful of spikes, perhaps a dozen, and a few pounds of nails, sixteen penny nails, eightpenny, and sixpenny. Back at the first cache, resting in a stand of small spruce, about one hundred twenty-five miles from the cabin, five rolls of green asphalt roofing and some boards waited for another year.

We built the cabin well, but not close to perfect, of course. Charles and I considered that for years. Lots of things to do differently. But the imperfections made the outcome special and ours. Things are good but you're not sure how good because they are different from what you would do if you did it again.

I couldn't have finished the cabin without Charles. He was always so determined and capable. When we lived in that little town on the Oregon coast, down near the bay, Charles, in high school, wanted to fish out on the open water, so he built a boat with a few dollars and a book, not a rowboat, but rather a twenty-footer with a cabin. When I think of small bay skiffs, I see Charles's. He and I, our sisters, boats, bays, rivers and shiny fish are linked forever, as are all of us, of course, linked by space, blood, and water. All of us on the earth, squirrels and trees and man, and all the rest, share a past.

All the water in the world is all the water there has ever been. It just travels endlessly about. One day after the trip, I poured a tall glass of well water from the tap at my house in town. I wondered if any of that water had been drunk before me. Possible? Probably not, it turns out, but it is likely my glass of water and all the others across the world was once dinosaur pee.

9

Basking in River Shallows

"We should go to Oregon this summer and see family," Doylanne said. We sat at the kitchen table. Jake sat on her lap eating cereal. I drank coffee and read about bear attacks.

"We can't afford the airfare for six and go to the cabin too. There's work to do on the cabin." I didn't want to go to Oregon. Why go somewhere and be hot?

"Boys haven't seen family in a while."

"The road goes both ways. I don't have a stove installed yet, and the roofing's not done."

"You said it was covered."

"Roofing paper's all."

"You just don't want to go to Oregon."

"I didn't say that."

"You were thinking it."

A friend and I fell short in our attempt to reach the cabin in June 1990. At about river mile one hundred and thirty, the head gasket blew on the outboard motor. The outboard sounded like playing cards on bicycle spokes. We weren't going any farther. We poled over to the shore, and in the company of mosquitoes, replaced the gasket. The adhesive used on the gasket called for a twenty-four-hour curing period before running the outboard, so we were there to stay for a while. I walked back through the forest looking for a meadow, but the terrain rose to a ridge behind us and I turned back to camp near dark. That night on the bank we sat close to the small campfire and talked about past hunting trips, big salmon, and the circuitous route to our camp along the river.

I often felt tiny along the river. The vastness was obvious, but it wasn't until I stopped moving for a moment that I felt my true size. Carl Sagan wrote, "We are like butterflies who flutter for a day and think it is forever." I was at times consumed with self-importance, but the woods blunted that fallacy. It is hard to overstate the triviality of a single man in the wilderness here on what Sagan described as a pale blue dot in the universe.

As the morning white light swept over us, we used the pole to reach the current and drift away from the bugs. We drifted several easy hours and "Let her float wherever the current wanted her to," to borrow from Mark Twain, before firing up and heading home.

In town, I talked with a machinist, hands like rusted wrenches, who said my gasoline-oil mixture was likely too lean, so the outboard ran too hot, warping the head, ruining the gasket. He removed the warp by shaving the head, a neat machinist's trick resulting in increased compression and a more powerful engine.

I'm indebted to my friend for his practicum in river-bank small engine repair. Over the years, I have become competent tinkering with small engines.

Back at home after the aborted trip, I prepared for my first school principal job at a beautiful little elementary school. I had some meetings to attend at the district office and some teacher interviews to conduct, but what was on my mind was taking my two oldest sons, Jack, 12, and James, 10, to the country. I now had some time on the rivers and knew a little more about outboard motors. That was good, but the distance, remoteness, and lack of communications worried me. I'd made some bad decisions, too. I constantly told the boys and their mother of all the horrible things that could happen.

Fall overboard, you drown
Get lost in the forest, you're lost forever
Get shot, you're dead
Run into a bear, you're eaten
Get a splinter, you die of infection
Cut by the chainsaw, your leg is hamburger

It was all gruesome and our time on earth was growing short.

"Don't do that, Eric. Don't scare everyone." One boy rested on her hip and three hid behind her. My young sons were about to get in a little boat and travel hundreds of miles into the wilderness with me.

In early July, the boys and I left for the cabin. Jack, coat collar zipped up under his chin, sat in the seat by me and James, a wolf hat pulled over his ears, made a seat in the supplies as the boat pulled out onto the river, but they soon crawled under a tarp. Cold ruled the river that day in July. Occasionally, one wiggled free and searched for something to eat and would ask to run the boat. They took turns doing this. They both ran the boat some. At about mile one hundred, with smoke from a forest fire visible to the south, I picked a camping spot along a high bank with tall, straight spruce, highbush cranberry bushes and an expansive view of the river and adjacent hills.

Jack and James started a fire while I set up the tent and built the camp. James chased a frog. The frog hopped to the river's edge, and James followed, tumbling into the river. I lunged to the edge, and he pivoted and scrambled back onto the bank.

Day one.

He changed into dry clothing, and Jack added dead branches to the fire. I played guitar, running up and down a scale trying to make it bluesy, while the boys discussed swimming and tree forts.

In the morning, we bundled in all our clothes and James ran the boat into the smoky air. We took the chilly morning face on, waiting for the sun to clear the spruce tops. After a bit, the boys disappeared under the tarp, and I ran deeper into the smoke.

Like most years, forests fires burned across the state. Thousands of lightning strikes slammed the interior, some resulting in fires that burned until the rainy season in August. During the summer of this trip, according to the forest service, nine hundred and thirty-two fires burned more than three million acres in Alaska, about the same as Connecticut burning. Smoke dangled at the river's surface all day as we ran into the yellow haze.

In the afternoon of the second day, we rounded a corner, and saw a man in a skiff crossing the river. We surprised each other. He made the universal hand signal for drinking coffee, so we followed him to the bank. We jumped out of the boat and followed him up a steep bank to his cabin hidden two hundred feet back in the woods. His old cabin stood about five feet high at the ridge pole. We stepped down two feet when we entered. He made coffee for me and Tang for the boys. We talked for an hour. Friendly, helpful man, he encouraged us on our way that day. He was a trapper living full time in the woods.

We waved goodbye and kept on. Later that evening, Jack held the boat against the bank at the homestead. We scampered up the bank, and there sat the little log cabin built the summer before. The boys stayed close, and we circled the cabin and counted the bear claw marks. The acrid smell of wood smoke burned my nostrils.

The next morning, we shed clothes as we unloaded. The mosquitoes hid in the brush away from the direct sunlight, but the horseflies, the ones with wraparound sunglasses, thumbnail size, dive bombed and attacked. Like mosquitoes, the females bite. They loved James.

Beaches crept out half way across the river on most corners. The boys eyed the sandbar at the first corner upstream. By midday, Jack and I began throwing camping gear back in the boat to run up to the beach. We'd had enough of the bugs. Where was James? We called out and my pace quickened until I heard him in the cabin. He sat on the floor in a dark corner, refusing to come outside.

"What're you doing, James? We're going to camp up on the beach at the corner. It's time to go swimming."

"They got my arms," he emerged in the doorway.

Welts the size of dimes covered his arms and neck. James was having an allergic reaction to the horseflies, or at least that's what I thought. That worried me. I'd once experienced an extreme allergic reaction.

About twelve years-old, I slogged out onto an Oregon bay at low tide with a shrimp gun, a cylinder that is forced into the sand, and the air vent hole on the top blocked with a finger or thumb. When the cylinder is pulled up and the air hole unblocked, the sand falls to the ground. I was mining for sand shrimp to sell for bait down at the docks. When my bucket was full, I headed to our apartment a block away. No adults were home. I took a handful of the shrimp and tossed them in a pan of boiling water. I ate the tails. Within thirty minutes my head looked like a basketball.

Horseflies attack in the bright sun, so James was free of them in the cabin

"Let's get to the beach. It'll be better there," I told James. I could see him considering that. He was serious about staying put in the dark cabin corner, and it took several minutes for Jack and me to convince him swimming was the answer.

I set up camp. The boys played in the river until after midnight. In the shallows, just before the channel, the water was warm, and James rolled in the sand, horse flies forgotten, at least for a time.

"We're going to stay in the cabin tonight" I told the boys the next morning. You think Mom will like it here?" I asked. "She won't like the mosquitos or horse flies or how far it is," James said.

That was a fast response. We ate oatmeal and drank hot chocolate and talked about swimming in the afternoon.

"She'll like the cabin," Jack said, "if we can keep the bugs out."

We ran to the cabin and chinked it with fiberglass insulation and covered the window openings with clear plastic. A bad idea. I considered leaving the windows open but thought hot was better than buggy. Wrong. Mosquitoes don't like sun, but they don't seem to mind hot. We spent a miserable night in the cabin. The mosquitoes flew and landed like the place was theirs.

I'd forgotten to pack insect repellent coils. The boys and I devised a variation that worked, somewhat. I filled a pan with leaves and cooking oil and set the concoction on fire. The smoke kept the mosquitoes lying low, but our little brush fire had to be kept alive. We experimented with birch and willow leaves but couldn't determine which was the best insect deterrent. I restarted the smoldering leaves several times during the night. Once, while rekindling, I heard steps on the deck. A large black bear crossed over and moved on and disappeared. I watched the boys scratch their faces and arms. The next day we moved back to the beach for the remainder of the trip.

Each day we cut and hauled brush in the morning, and in the afternoon, we played in the outdoors. Often James, in long sleeves, sat in the boat reading about Narnia. After a bit, he hunted. One day he killed a red squirrel with his BB gun. He hung the squirrel from a birch branch and there cleaned his first animal at the cabin. He rinsed it clean in the river and ate it with macaroni and cheese.

Smoke filled the land in the mornings, but it thinned by afternoon. We nearly left for home one morning because of the smoke. A drum of gasoline sat in the boat and the smoke covered us. I imagined wayward sparks igniting the gasoline and blowing us all to bits. But, again, by early afternoon, the smoke cleared.

One evening, late in the trip, the boys sat on the cabin roof with a rifle. Jack wanted to hunt bears. His hair had grown, and his face had taken the color of the cinnamon beach we camped upon. I sat in the cabin and listened to their low voices and soft movements on the roof and hoped no bear crossed the yard. A few hours later we putted back to the beach.

We loved it all deep in the woods because it felt real. What we did mattered. Unfeigned experiences in town are rare, but in the wilderness, they find us. We grow when afforded the opportunity to find solutions.

Benjamin Bloom, education scholar, wrote that understanding moves along from simplistic to complex. Simple: Where are the beans? Did you remember to put in the drain plug? How many gallons of gasoline do we have? More complex: The deep, circular gurgles and eddies mean deep water. A boat on-step will go faster and use less fuel. More complex: Use the engine to hold the bow to the bank so your partner doesn't fall in the river. Putting the firewood deeper in the woodstove may improve the draft. More complex: Rain in the mountains means the river will likely rise. A wounded moose will lie and die if not agitated. A teenager will want to go home early if he has a girlfriend. Most complex: It's too late to shoot.

Exposure to that progression is a treasure offered for free in the outdoors.

Extended time in the wilderness demands introspection and the understanding of connectedness. John Muir wrote in *My First Summer in the Sierra*, "When we try to pick out anything by itself, we find it hitched to everything else in the Universe," a universal lesson easily taught in nature. The connections can be seen and when beyond our ability to fathom, can be imagined. In the outdoors, a full day can be consumed doing a simple chore and a full understanding of the chore must be mastered to finish and not waste materials and not get hurt, a valuable lesson for all of us. Nature provides the workshop. Kids must do things outdoors regularly to become engaged with nature. It's not enough to witness from a distance. Visits to national parks have declined in recent years, maybe because some of the fun is gone. A trip into nature should not be a trip to the museum. Look but don't touch won't result in lasting connections. Sitting under a tree may be enough, though. The closer the better. There doesn't need to be an outcome.

We shouldn't make kids choose between nature and technology. Nature will lose. Lewis and Clark would have used a GPS if they'd had one (so would have I). Use technology to connect youth with the outdoors. Send them off on a journey with a compass. Perhaps the most natural connection between nature and technology today is the development of renewable resources. Have them maintain a trail of solar lights in the backyard. Ideas are everywhere. Talk with children about ecology. Words matter. Most people don't have to experience something to understand. That's what our brain does for us and makes us intuitive human beings. Teach children that nature is not what you can take from it. An experience in nature doesn't have to mean a fish or a moose. A square foot of soil with worms on an urban deck can teach a lot. Worms, after all, help to increase the amount of air and water that gets into the soil. Find a local cause. There will be one. You don't have to march but you can express your opinions to your children, and most importantly, find a place to take your children that they can learn to love.

Richard Louv, in his 2005 book, *Last Child in the Woods*, writes in depth about the concepts mentioned above, and adds a disturbing conclusion: We are all spending less time in the outdoors, particularly children, and the result may be a wide range of child development problems.

It was time for us to go home. We packed a much lighter boat for the run back to the launch. Another hot day, both boys took a bath in the river. James seemed more at ease with the horseflies. When one buzzed by, he went on a search and destroy mission. Jack wanted to hunt for bears during the journey home. I told him to keep his eyes open.

On our way home, with Jack running the boat, I fell asleep. I awoke more than an hour later with James at the helm. We moved along smoothly, on-step, and in the right direction. They both beamed. I closed my eyes and slept

some more. The river journey home took a long time; it always does, but there's a value to traveling a route that constantly changes. The decisions are incalculable, and some of them are life defining. During those two weeks at the homestead the boys grew and learned a lot about the river and the woods, so did I.

When we pulled into the driveway, Doylanne ran out to meet us. She hugged the boys and checked the length of their dirty hair. Jack said, "Mom, you got to go there with Dad."

"Yes Mom," said James.

That night I checked on our little garden. Doylanne brought me a beer as I cleared weeds. We talked about the boys and agreed that a child's head isn't empty, needing only to be filled with knowledge, as some people seem to believe. Children, time and again, at a young age, ask questions about nature, the universe and life, far beyond the information taught to them. They learn by both watching and doing. The more playing and experiencing, the more learning. This is all easy and natural in nature.

"Congratulations, Eric. You now have a cabin." She hugged me.

"It's your turn next," I said.

10

A dead end is only a sign

Wilderness bursts with romance. Swans mate for life. They lie in ponds, necks stretched across their bodies, watching. Moose call provocatively across the forest, tundra and taiga. Mice sing to their romantic partners while trees wrap around one another, a permanent clinch, spiraling upward.

We were not alone.

Four years after the staking trip, Doylanne rode with me to the cabin. For the first time in their lives, the boys were not with one of us. I wasn't worried about the boys. They were with good friends. I was worried Doylanne wouldn't like it. Maybe none of it.

Doylanne, bundled in three layers, pulled her hair under a stocking cap. Peach nails, no lipstick. The boat rose on-step easily with our light load, and I was chilled right away. I piled up gear in front of us and several times cut the motor and drifted so we could warm up some. The weather broke nice about two hours in, clouds flying before the sun revealing the mystery we wait for, and the temperature quickly rose. Still not warm, by any means, but better. I immediately gained confidence.

"Okay?" I asked. Her cheeks were red and looked cold but she smiled. "If you get cold, drop down behind the gear."

I met Doylanne a few days before we began our senior year in high school. A friend was taking me home from morning football practice, and we pulled through the turnaround at the end of main street down near the bay in a little Oregon town

I would later learn she and a friend stopped to prolong their afternoon after a school shopping trip. I recognized her from school but had never met her. I noticed she had a book on her lap, and I'll linger here for a moment, visualizing the bay with its wooden boats, the smell of the docks, the sight of the gulls circling about overhead, the hillsides of Douglas fir, and the small-town main street.

I must have looked rough. My coach had just tried to kill me with wind sprints and a human torture technique he called deer runs. A deer run involved lining the team in single file fastest to slowest then sending each of us off on a six-hundred-yard sprint, releasing us in five second intervals, each person trying to run down and pass the person ahead of them. Ten more sprints for every time passed. I didn't handle that so well. Sometimes I'd have fifty sprints extra.

Doylanne, though, looked beautiful: long, perfect hair, stunning brown eyes, a smile that made me smile. Smelled good too. When I think of this hot, late August day, another football practice scheduled later that afternoon that I wanted to skip but knew I wouldn't, I recall the uncertainty of life at that time. I knew there wasn't much time left for doing things that mattered little, at least that's what I was told and what I believed. I didn't, though, have much of a clue what that meant. Doylanne and I talked for a few minutes, neither of us knowing, of course, that this chance encounter near the docks would impact the rest of our lives. I got a phone number and called her that night. I've talked with her almost every night since then, forty-eight years ago.

Four hours into the trip, about eighty miles of river, Doylanne and I stopped to camp. As we unloaded our camp gear, our eyes followed flocks of passing Canadian geese and noisy Sandhill cranes. The fall temperatures calmed the mosquitoes, and with a slight breeze, they were scarce at the campsite. Across the river, an extraordinary display of small yellow flowers mixed with rose hips lined the bank. As it grew dark, Doylanne stayed close by the large campfire thoughtfully arranging sticks at the edge of the fire. I pulled on a knit hat and put up the tent, fluffed up the sleeping bags and brushed out the sand.

Doylanne called to me. "Come here. Do you hear that?"

Into the fire light reaching out several feet onto the river surface swam an anvil head beaver wailing nnnng nnnng, the sound we might make if we closed our lips and breathed out our nose and forced a sound from our chest. Maybe.

"Sounds like singing," Doylanne said.

"Yeah."

Another beaver splashed just out of the firelight. The singing beaver circled within our view and faded into the darkness.

"Beavers mate for life," I told Doylanne. "There are Indian legends about singing beavers." I'd read a book about it, I told her.

"It sounds like a cello," she said.

"Both parents take care of their pups and older pups baby-sit their little brothers and sisters."

Doylanne turned her attention back to the fire. She poked at the coals. "Will the fire burn most of the night?" she asked, studying the fire.

"Are you getting sleepy? We should try to get a good start tomorrow."

She tossed a small log on the fire.

"As late as it is, there will still be hot coals in the morning," I told her. "Doylanne, the thing about beavers is they won't come near the shore if a bear is around."

She smiled.

"You seem to know a lot about beavers."

"If a beaver swims upstream on its back, there will be an early winter." A beaver tail smacked the water not far up stream. Another smile.

I stood guard by the tent as Doylanne slipped into her sleeping bag.

I ate a beaver once. In the majestic Kilbuck Mountains in southwest Alaska, Charles and I camped on a gravel beach in a gorgeous spot in nowhere, a fishing trip into the mountains, searching for enormous rainbow trout. Our discussion turned to mountain men and food. I'd read beavertail was thought to be a delicacy by the old timers. I shot one minutes later, and we boiled the tail and roasted the beaver over an open fire. We jabbed at the tail in the boiling river water and agreed we wanted it to be well done. We sliced through its tough skin to find an off-white grainy substance the consistency of thick peanut butter. We spread it on pilot crackers. It was fine. We ate both the beaver and the tail.

Doylanne and I left early the next morning into a cold breeze, but the sky promised some heat. At about the one hundred fifty-mile mark, the river petered out to a depth of inches and became impassable. This surprised me because just weeks before this stretch had plenty of water. I couldn't find a place to go. I drifted back and tried to find a channel, again and again. I threw the anchor out to hold our place and stood on the bow trying to understand why the navigable channel disappeared. I couldn't discern the bottom through the glacial silt, so made running decisions by the corners, banks, visible sandbars, drifting wood, and the surface. Nothing drifted. I couldn't read the river.

I stepped out of the boat in hip waders, and Doylanne moved to the middle of the boat. I slung the bow rope over my shoulder and pulled. Not much progress. I moved a dozen feet ahead of the boat searching the bottom for more water. I pulled until the boat stalled, then trudged back to the boat and shoved it free. When it floated, I tried different angles. Doylanne pushed with the pole.

There was nowhere to go.

After two hours I gave up, and we drifted back a short distance while I guided us toward a beach. I set up the tent and my head pulsated, hurting like two days without water. I poked my thumbs into the cervical vertebrae. Doylanne searched for firewood. There was nothing to unload on the beach to lighten the load, nothing of consequential weight besides me, Doylanne and the fuel.

"We'll try again in the morning. Maybe the water will come up some," I told Doylanne.

"It's okay, Eric, if we don't make it. Maybe we shouldn't make it."

We sat near the fire until late and watched the gray water become black and the pink sky transform to stars. I heard Doylanne fall off to sleep, but sleep was not for me. I sat by the fire all night. I felt that the river was keeping me in my place, holding me back, but that's not what rivers do. Must be me. It was time to do something. The water didn't rise by morning. I ran us back to where the water was too shallow to navigate, and we both got out of the boat and pushed further on to the sand hoping to break over into deeper water.

"What if we can't get out," Doylanne asked.

I took the bow rope and extended it another fifty feet and slid out in front of the boat and the water depth didn't improve.

"Okay, let's go back and think more about this."

I built another fire and made some coffee. I again forced my thumbs into the base of my neck. What would cause this to happen? The water to disappear?

"Should we go home?" Doylanne asked.

"I have another thing to try," I told Doylanne. I'd hit a dead end on the river, but a dead end doesn't mean stop.

Doylanne and I pushed off from the beach, headed downstream. A short distance down was a braid of river. I always passed on this route as too shallow. I turned the boat pointed upstream and held steady in the current for a few minutes thinking about this choice. The branch, going east, ran along a ridge. But I wasn't hopeful about getting the boat through the entrance of the fork. I could see branches lodged in the bottom, and the water surface held a tight ripple indicative of shallow water. There was, however, a negligible sandbar with a sharp cut, a sign of a little more water.

"Doylanne, get down and hang on. If we hit, there'll be a good jolt. Hang on!"

I accelerated onto step. Doylanne knelt between the two bench seats. The boat slipped over the threshold of the fork along the edge of the sandbar about twenty miles per hour without touching. Within seconds we sailed in substantial water. I held the boat on-step and against the high bank, and we re-entered the main river and two long corners were removed from the trip. It turned out to be a short cut. I used this route for a few years before the entrance became too shallow and a channel in the main river became passable again.

There are three and a half million miles of river in the United States, and we had about fifty of those to go.

We arrived at the cabin in late afternoon. I ran the boat against the bank, holding it there so Doylanne could step out with the rope.

I followed with the rifle and grabbed the rope from her to tie off to a small tree. The bank was in full bloom, but the rose petals had given way to rose hips. Doylanne stayed close and we hiked up the short trail together.

It can happen most anywhere: the lighting of a Christmas tree, a white horizon, glassy ocean, shining faces around a birthday cake, or glistening rain under a city street light, those rare moments when the setting is perfect. Here the temperature was nearly sixty degrees with lethargic horse flies and mosquitoes. The afternoon sun strained through the spruce and a dash of light crossed the cabin roof and faded in the brush.

Doylanne stopped. "Beautiful Eric. It's a gingerbread house."

We walked around the cabin and I pointed out details in the construction. Ants and pitch fashioned a fascinating mosaic along the full length of several logs. We carried the gear into the cabin, and I stapled screening material over the window openings and lighted an insect repellent coil.

It seemed perfect. Doylanne and I spent our days building a log bed frame, a dining table, and a railing for the porch. We designed a floral arrangement featuring autumn shrubbery for a centerpiece and used our best cabin dishes.

We completed little chores, but most of our time was spent sitting on the river bank. Slow and easy. The fly fishing was good, and Doylanne helped me sight in the rifles. We hiked the property line together clearing wind-blown debris and cutting back new growth of willows and birch. Trumpeter swans beat their wings and sounded off on a small lake near the cabin. Sounds of the beating wings resembled the echo of hammer on wood at a distance: bamcha, bamcha, bamcha, bamcha. Swans splashed onto the river and stayed in view for hours. Sandhills in enormous flocks passed over the cabin. One evening a young bull moose with maybe a thirty-inch antler spread stepped out on a beach across from the cabin. We watched from lawn chairs at the edge of the river. The moose, dark, standing in the shadows, took his time. He drank from the river, submerging his snout. He lifted his nose to the air, upstream, downstream, and drank more. After several minutes, the moose ambled back in the brush.

* * *

Years later, in mid-September, James, a Medievalist at Cambridge University, leaned against a birch tree watching for moose. He held a hunting rifle across his legs and slapped at the occasional mosquito. He scanned Jake Lake with binoculars. I did the same one tree away. Cold moist evening, rain felt likely, maybe snow.

We watched the meadow for a bull moose to show himself. Time for the rut, a moose might step out in the open at any moment. The rut refers to when moose breed. This happens in the fall from late August to early winter. This is also when the female moose comes into estrus. In August and early September, male moose scrape the velvet off their antlers, and the banging and scraping of antler against trees can sometimes identify the moose's location. As the rut becomes more intense, moose, both bulls and cows, lose their natural cautiousness. They can sometimes be called out into the open by imitating the sounds of scraping antlers or audible calls that sound like a moaning cow or a determined bull.

The peak months for the rut in Alaska are October and November. This is also when the leaves are gone and visibility in the forest is the best. So, for the sake of moose, almost all hunting seasons end before October. Typically, the hunting season in Alaska is from the first of September through the third week. The rut is often underway, but not intense, and the leaves, though turned to autumn colors, haven't fallen yet.

We watched Jake Lake, an orderly and symmetrical drop of reflecting light, a refuge for waterfowl, with forest on three sides with an outlet at one end. A beaver dam stood prominently in the center.

A pair of trumpeter swans departed from the lake. They slapped their wings on the water during their long take-off run, pulled in their necks slightly at lift-off, and pounded into the sky. They extended their necks fully and disappeared over the trees. Maybe they were spooked by a woodpecker near us making a racket, or maybe us. Their two cygnets, still gray, were in the lake. The swans circled twice, taking their time. They blew their trumpets in a deep, resonant call as they passed overhead.

Trumpeter Swans travel from the western United States to Alaska in the spring to breed. They live near the cabin throughout the summer and fall. They seem to change colors with the changing light and are without doubt the stage stars of waterfowl.

The landscape where James and I sat is shaped by freezing and thawing over unfathomable time. Yet it is the calculation of minutes that occupies our thoughts. Knowing how long it takes to walk to the river can save a river run in the dark and prevent a collision with a log. A decision made at 6:00 p.m. affects the experience at 9:00 p.m.

James used the paper birch, more than fifty years old and only six inches in diameter, as a back scratcher. The adult swans, gone too long, sailed in for a landing, and the swan family headed single file to the lower end of the lake as the good light vanished.

The forest floor was covered with spruce needles, inches of them fallen through the years. James and I were engulfed in spruce and inhaled them with each breath. I sat, crowded by spruce, and became saturated with the smell. Will the aroma of freshly baked bread lose its appeal if all we smell is freshly baked bread? I stood and faced the breeze and allowed the passageways to be swept clean. I sat back down again and breathed deeply and checked the time.

James's boots rested on a rose bush. The flowers were long gone but the rose hips were ripe. He brushed leaves from his coat. Red squirrels bantered a few trees up the ridge. James stood and leaned against the tree and became more intent with the binoculars, probing for movement and dark shapes just inside the foliage.

Binoculars allowed us to venture into the trees. Late in the season is best for this because leaves have fallen. The highest and driest ground at a meadow is where the tall bronzed grass meets the forest rim. Moose beds are often found there. Moose travel back and forth between these two environs. They can stand out in a breeze to deter bugs or go down to the small pond common in these meadows for a drink. They also can retreat rapidly into the trees. Most of the moose I've seen out here through the years have been on the edge of a meadow.

A calf sidled out to the edge of the pond. James nudged me to make sure I saw the moose. The calf led the way before the cow showed herself. No bull. We waited for a bull.

How big are bull moose? A world record moose with an eighty-inch antler spread was killed during a hunt near Palmer, Alaska, in 2017. A moose ten-years old or so can grow antlers in this range. At least one of them has. An average adult Alaskan bull moose, though, lives with an antler spread between forty-eight and sixty inches. It takes a moose more than six-years to grow antlers in this range. The size we want is considerably smaller still, about thirty or forty inches, a young bull just a few years old. There are more than one hundred seventy-five thousand moose in Alaska. We intently watched the clearing for just one bull to appear. James pointed to some spruce trees to our left. An owl was calling from there. There was still time.

We continued to follow the cow and calf. The big cow, probably more than a thousand pounds, watched her calf and occasionally turned to look back to the woods. Of course, like us, moose don't know the time or the hour of the last breath. For me, it was time to exhale, count the seconds as air escaped, slowly clear my lungs. It was time to stamp an image of this place in memory, to experience the cool temperatures and fading light and appreciate how special this all was.

I left James and slipped down the hill to the open end of Jake Lake. I picked my way as softly as I could but I'm not light footed. Twigs snapped. When I reached the bottom of the ridge, I heard James make a cow call. I couldn't see him. He did it again. I edged down a few steps toward the lake.

A bull stood ten feet out from the shore, probably fifty yards from me, its snout pointed at James up the ridge. I aimed behind the front quarter and fired. The bull toppled over into the water. It was dead when I got to it a few minutes later.

It'd be dark soon, and we had a bull down in a foot of water. James joined me in minutes. Nice fifty-inch plus bull. We left for the boat to run back to the cabin for a generator and lights.

Jake Lake is named after Jake, of course. Years before, all of us were hiking along a ridge off the river one afternoon when Jake, six years-old, called out, "There's a beaver house." He dashed ahead. We christened the pond Jake Lake. I imagine that Jake has met most of the people who have ever walked around this pond, maybe all of them.

We headed back to the boat, James leading the way up the ridge. Dark under the trees and no trail, we stepped over fallen trees and picked our way through the underbrush. The hiking was fine except for a short span near the river where James crossed a narrow swamp, and I followed in his tracks.

Skylight outlined the river banks, and the river curled to black. The twenty-one-foot bay runner with dual fifty horsepower four-stroke outboards jumped on step. James throttled down and trimmed the props to as shallow as possible, and we cruised downstream to the cabin. We loaded gear to handle the moose in the dark. Two hours later, James and I pulled the moose out of the water with a come-a-long, a manual winch with a drum and a ratchet. We used lights powered by a two thousand-watt generator and fifty feet of half inch nylon rope. The moose pulled easily through the water, then with considerable effort onto dry ground. Another hour later, the moose was gutted and covered with a tarp, and we headed back. It would take us all the next day to hang the moose meat at the cabin.

We carried out about seven hundred pounds of meat and sixty pounds of antlers.

Hunting moose is a big deal in Alaska. More than twenty thousand hunters in general harvest hunts seek Alaska's moose each year, and more than four thousand of them down a bull. A general harvest hunt is for all of us who buy a license and just go hunting. There are other moose hunting opportunities such as subsistence hunts through an application process and drawing hunts for a fee per chance where another twenty-five hundred moose are killed. Most hunters, though, participate in the general harvest hunt and the chance of filling the freezer is about twenty percent.

I've probably hunted twenty of the thirty years I've traveled to the cabin. Sometimes I get a moose, but I've never counted on it. Moose are usually hard to find. I'll go up for weeks and come home empty, and I'll have friends who get them on a weekend close to home. I didn't stake the land for moose, a good thing. Remote doesn't equate to ideal moose habitat or easy hunting. Our homestead offers neither of those. It's a long way to go to maybe run into a moose.

But sometimes they'd find us.

Thoreau wrote: "At a certain season of our life we are accustomed to consider every spot as the possible site of a house." Doylanne and I sat at such a spot a few years after my hunting trip with James. We were searching for berries and moose. We sat on a hillside where we'd found blue berries and dreamed aloud. We ate our sandwiches and sipped soda pop, and picked maybe a pint of berries while watching the light clouds sail past the peaks. By late afternoon, after a lazy day resting against a grove of diamond willows, and whimsical planning of cabins, it was time to move closer to the river to get near the boat.

We stopped at the edge of a spruce forest. Ahead of us was a meadow of chest high grass like a farmer's unkempt field. We saw a narrow pond, reeds, swamp, and sandy colored grass to where the meadow rose to meet the forest on the far side. Here we picked low bush cranberries, and I tried some moose calls. Doylanne ran her fingers through the bushes, yielding pearl sized berries. We filled a gallon container in an hour.

I glassed the meadow with the chest high grass while Doylanne picked. Often, she stopped to watch. I called for moose, scraping with a piece of antler. I swept the antler lightly across branches and listened. I did this a few times and went back to berry picking. Half an hour later, I lightly smacked the antler against the tree. I wanted it soft. During the rut, bull moose challenge one another to determine dominance. Small bulls don't have a chance against the big bulls, so they stay out of the way. A wallop too loud might keep the bull I'm looking for in hiding. But I grow impatient. After an hour, I gave the tree a good whack, a fighting call. Come and see who's the toughest. Seconds later a crash across the meadow startled us. Ducks in the pond squawked and hit the air. Doylanne spotted the bull first. It was much too far away for a good shot. We planned in low voices. She would stay put; I would move closer.

I inched along the edge of the meadow to make up distance while the bull hung at the edge, his head down. I had a place in mind to take a shot. I stepped back into the trees and snuck down the meadow slowly. My coat scraped branches, and I stepped on dry dead leaves. I tried to step like a moose, walk like I had four legs. My version: I took a couple soft steps and waited. Repeat.

I came to a break in the trees, scanned with binoculars but couldn't see the moose. I looked for Doylanne and couldn't see her either. I sat back, pulled on a stocking cap and tried to get my pulse down. The odds were good the bull was close, standing back in the trees or lying down in the grass.

A big bull appeared before me across the meadow. A cow grazed beside the bull. I moved a few feet down the meadow to look for Doylanne. I found her pink knit stocking cap at the top of the grass. I took my time placing the crosshairs on the moose, just behind the front quarter. I squeezed the trigger lightly. Boom. The moose jumped ahead but didn't fall. Before I could take a second shot, it stepped out of sight. The cow disappeared too. I watched for some movement inside the trees. Nothing. I sat back down.

My pulse slowed as I swiped at mosquitoes. I cleaned my scope and glasses and watched where the moose disappeared. A hawk circled near where Doylanne would be standing in the grass.

It's presumptuous to contemplate the best way for anything to die, but I think about it when I'm hunting. I don't like killing animals, but a well-placed bullet must be the fastest way for a moose to die, therefore, it seems to me it's the best way. Moose die lots of ways, like all animals. Predation is common, but moose also die from disease and malnutrition. Few, if any, moose (or bears or squirrels) live to an old age and die peacefully in their sleep. At some point, they face a difficult death. If I take care of the meat and eat it, then I can live with a well-placed shot that kills the moose rapidly.

All hunters should be prepared to make a good shot. An animal shot full of holes is stressed before it dies, and the meat can be affected. Livestock scientists, most notably Temple Grandin, livestock expert and Professor of Animal Science at Colorado State University, have found at least two unsavory conditions resulting from pre-death stress. If there is lactic acid buildup before death, caused by high stress, the affected meat is pale and soft. Equally as disturbing, if the animal lacks glycogen, a condition caused by stress, the pH may not drop rapidly enough after death because of a lack of lactic acid. In this case the meat will be dry and dark in color and maybe more likely to spoil since it lacks the lactic acid needed to slow the growth of microorganisms.

I hoped for a dead moose inside the trees. A moose, if hit hard will lie down and die, but if it's aroused, it can run and not be found. I worried about losing the moose. Aldo Leopold wrote in *The Sand County Almanac*:

> A peculiar virtue in wildlife ethics is that the hunter ordinarily has no gallery to applaud or disapprove of his conduct. Whatever his acts, they are dictated by his own conscience, rather than by a mob of onlookers. It is difficult to exaggerate the importance of this fact.

I slid quietly back to Doylanne. We talked, hugged and settled on a plan. Doylanne would hike back and stand by the boat. I would know where she was, at the opening in the trees where we climbed up the bank from the boat. I left the berry patch and stole a path around the meadow, picking my way slowly, taking my time to reach the spot where the moose disappeared in the brush. The foliage was sprayed red with blood. I stopped and located Doylanne's pink hat. I stepped into the trees a few steps and stopped. I closed my eyes momentarily to concentrate on sound then crept a few more feet. A twig snapped. Not me. I couldn't see the moose, but I could clearly hear sweeping through the brush and then sloshing in water.

I knew where it was.

I stepped out at the edge of the meadow, and the moose was wading across the narrow pond, stomping directly to where I took the shot. The moose climbed out of the water heading toward the trees. I didn't have a target except his hindquarters. I called, "Huuut!" The moose looked back. I placed the crosshairs at the base of the skull and fired.

Yes, that's what I called. It's all I could think of. Football fans know the sound I'm talking about. The quarterback breaks the huddle. Signals are barked up and down the line of scrimmage. "Omaha. Right. 23. 23. Huuut!"

I wasn't happy about the first shot. The second shot was nice.

We found the moose dead in the high grass laying spread-eagle on the ground. The good-sized moose had to be turned on its side to gut. I field dressed the moose like I would a small black-tail deer. Doylanne pulled on one hind leg while I pulled the other underneath. Progress in inches. We got the moose on its side. Doylanne held heavy, spindly moose legs in the air while I removed the entrails.

We gutted, skinned, and cut the moose in quarters. I cut off its head, so we could get all the neck meat. Once we removed the backstrap, we split the rib cage down the back bone and put all the parts into game bags. I sawed the antlers from the head with a chainsaw.

So much meat and so little strength to carry it. A quarter of moose is heavy. Skinned front quarters weigh about one hundred twenty pounds and hind quarters one hundred thirty pounds

The ground was relatively flat out to the river, and there was a clear opening through the high grass.

"Why don't we pull it on a tarp?" Doylanne asked. She took off for the boat and came back with a blue poly tarp. A quarter was laid on the tarp and pulled over the grass to the edge of the river. Eventually we carried the moose to the woodshed where each part was hung to cool and dry.

I like moose, the animal. Heavy bodied, fast, a drooping nose, and a dewlap under the chin, they are the largest of the deer family. They seem curious, friendly and they're so easily personified. I like them, and when I hunt them, I want them to die from one shot.

With a moose hanging at the cabin, we had time on our hands, the way I most like to be. I sat on the cabin deck that evening, after hanging the moose, and watched the meat, heavy, sway slightly as two mink pulled at the game bags soaked in blood. I ran them off with a BB gun.

* * *

The first trip to the cabin with Doylanne was amazing. In the evenings we ate grilled spruce hen with cranberry sauce while candle light frolicked on the logs. During the day we

spent most of our time at the river, but Doylanne loved a little pond a short distance from the cabin, just upstream and a quarter mile off the river. Ducks and geese cruised about, and a set of swans slid across the surface. We sat in a natural blind at the pond's edge with willows. A beaver hauled branches across the pond to a new creation. We waited for it to sing. It didn't. We named the pond King Beaver.

On an easy evening we talked about our trip in and her impressions. Experiencing the distance surprised her. She imagined miles and miles across a calm lake. She felt like I always did after turns and turns: a desire to just get there. Her sensory awareness also intensified, she thought. I felt the same way. In town, our senses are dulled by consistency and redundancy. We experience the same sounds, same temperature, same old smells. In the woods, what we see, hear, feel, and smell matters, but in town, not so much. In town, we can usually forget nature. Let it rain, who cares. This is the way we felt about nature's impact on day-to-day life in the early 1990s. Structures built to code, air conditioners and natural gas furnaces could keep nature in check. We could control floods and fires. We feel differently now, more than thirty years later. It's obvious that we can't really handle what nature can throw at us.

After eight days, Doylanne and I headed back to town to get the boys from our friends who helped make our trip possible. I had a new job to start, and we both were thinking a year ahead when all six of us would make the trip.

11

Home was a cabin too

"I think in two years we could be ready to move out there. Need to get a new snow machine. Maybe a couple rifles."

"That's it. That's all we need?" Doylanne asked.

"Maybe a generator. What else would we need?"

She turned away from James who sat in a chair in the kitchen in our little home. She was giving him a haircut. "Look around here, Eric, do you see anything we don't need?"

"We certainly could do without some of this. Who needs a television?"

"What about a refrigerator or a dentist? Eric!"

Doylanne turned back to James who was watching me and growing uneasy with his mother holding scissors. Time to move on to another topic.

"Jack'll be on the wrestling team," I said. He was in the seventh grade.

"They don't have one of those out there," Doylanne said.

I prospered my first year as a school principal. The teachers at this little school knew what they were doing. I gained an important insight that first year. Learning requires space. There must be time, for youth and adults, to reflect and self-correct. I learned that from the skilled, veteran teachers at that school. Success. At closing time on the last day of school, field day, of course, when the children loaded into cars and buses and went home for summer, you could hear the release of tension. I'll always love that school.

Now river time, and the journey to the cabin became significantly more complicated. The plan was that I would run to the cabin solo with supplies, in and out. After a couple days at home, Jack, 13; James, 11, and I would run up with another load of supplies in the boat, and Doylanne and Jed, 6; and Jake, 4, would fly in a chartered bush plane and meet us on another river about a dozen miles from the cabin.

I bought a new oil-injected fifty horsepower two-stroke and installed a steering column. We were moving up. On a warm early June morning, I pulled slowly away from our little cabin home near Wasilla to haul in the load of supplies to the homestead. Doylanne, with four children in my rear-view mirror, waved.

I'd stuffed the boat with food. Soda pop, box milk, evaporated milk, sugar, brown sugar, salt, pepper, bags of oatmeal, rice crackers, saltines, pilot crackers, gallons of peanut butter, white flour, sardines, canned ham, corn beef, smoked pepperoni sticks, rice, and hamburger mixes.

Six hours later, I motored away from the beach at the launch. At about mile sixty, I ran hard aground. I cut the power when I felt the beach at the bow but still slid upon the sandbar and ground to a stop. I rocked the boat from my seat. Nothing. I stepped off the boat and tied a rope connecting me with the stern. Water just above my ankles, I felt my way out in front of the boat a dozen feet. I extended the rope and edged out another twenty feet. Still no water. I tried out the sides. All sand. At the stern, I pushed. Half inch gain maybe. I had to turn back. I slowly swung the boat around. An hour later, the boat was turned upstream. I made inch gains, half inch gains and finally had to rest. I slept a few hours in the boat. No real darkness, for more than an hour during the wee hours I gazed out of a small opening in my sleeping bag at the clouds making faces at me and thought about ineptitude. Faulty sandbar navigation took hours, nearly every trip. Maybe I just went too fast. I idolized speed. Get there fast. Count the minutes. Measure success with time. I might say, "It was a great trip. It only took twelve hours." I needed to slow down.

I moved the gasoline in the morning. A mostly full fifty-five-gallon metal drum and two fifteen-gallon drums and three five-gallon cans ended up in the river. Most of the food was stored in two thirty-three-gallon garbage cans to keep it all dry. I wrestled garbage cans and all the soda pop over the side of the boat. I could move the boat now. I leaned hard into the boat, rested, shoved, rested, and finally broke free. I jumped in the boat and drifted below the supplies. Incredible sight. My supplies were in the river! I circled around and pulled up against the sandbar and reloaded the boat.

I made it to the cabin that evening, and near midnight with the sun still not below the trees, I went to the river to fish. For the first time on this river, I saw king salmon. They were blotchy red and certainly near their end. They entered freshwater at the mouth of the Yukon, hundreds of miles away.

In the morning, I headed home. Our home near Wasilla was also a cabin. We lived twelve miles out of town in an eight hundred square foot frame cabin on property financed with what the bank called a non-traditional mortgage which meant that the building was not up to par, but they'd give us a loan if we were silly enough to take it. The story and a half frame cabin had been moved to the property and set on pilings. The structure, built with two by fours, was usually cold, and plumbing was an afterthought. All the copper pipes ran in the open air under the floor and regularly froze during the winter. I spent so much time down there that Doylanne suggested I hang the boys' school work beneath the house so I would see it. I eventually skirted and insulated the exposed crawl space. The cabin took a lot of work but eventually did become a beautiful little home.

I rested a day then Jack, James, and I left home early in the morning and ran the entire trip to the homestead cabin, pulling in near midnight. We experienced a beautiful day on the water with the boys running the boat most of the way. I mostly watched the water and the boys and thought about how I would build a shed. I think I'm smarter on the river than on land. Maybe it's just being more attentive and decisive. There's no time to over think things on moving water.

Jack tied off to a birch and mosquitoes immediately attacked. We rushed to get the bear boards off the cabin door. Mosquitoes rampaged for several horrible minutes. In mouths, ears, and eyes, covering hands and necks, the mosquitoes made a lasting memory. It didn't go well with the nasty bear boards either. I struggled with the nails, dropped tools while swatting mosquitoes and my blood spattered against the house logs. Our bear boards consisted of dozens of sixteen penny nails pounded through plywood and nailed to the window frames and the door jamb. Pain inflictors. We pulled tufts of bear hair from the nails.

I tossed the bear boards off the deck, and we dashed into the cabin. I sat in a plastic lawn chair at the back of the cabin and let my legs go limp. I closed my eyes and I could smell flower perfume until Jack lit a repellent coil. My legs relaxed in minutes. I didn't want to move from the chair. I rested and the boys worked, carrying up from the boat what we had to have for the night.

Like the year before, we worked in the mornings and played in the afternoon heat. Our task was to get the cabin in the best shape possible for Mom and the little boys. The screening material Doylanne and I installed was still in place and we covered the doorway with an insect net from an army supply store.

Doylanne, Jed and Jake left home at 3:30 a.m. for the long drive to Fairbanks to catch an hour-long bush charter flight at 10:30 a.m. Our river at the cabin was too shallow for a plane to land, so the plan was for it to land on a river about an hour away. Jack and James and I were up early. We headed out at 10:00 a.m., planning to wait in the sun. After two river turns, though, the engine sputtered and quit. The boys kept the boat in the middle of the river using the push pole while I inspected the mostly new motor. It started, ran a bit, then died. I went through the troubleshooting list in my head. Nothing I could see. It ran, died, ran, died. I was worried we weren't going to make it on time. Would they fly back to Fairbanks? How long would they wait? I probed the motor like the mosquitoes did my head two nights earlier. I pulled the fuel filter off, and there it was. A clogged sink. How'd all that get in there? I cleaned the obstructed filter, and ten minutes later we were moving. We heard the plane, and we all pointed to the sky as it swung wide away from us, tipped its wings, and headed downstream. They waited for us on the beach.

"Where you guys staying?" the pilot asked. The little boys played in the gravel at the river's edge.

I studied the pilot for a moment. I didn't want to tell him, but it was a fair question.

"We have a place up there, up that river," I pointed to the south toward Denali

He shook his head. "Never stopped here before, don't come this way often." Wearing hip waders, he kicked the plane off the beach, fired up, veered to the middle and taxied upstream, before abruptly turning and taking off with the current. His plane lifted off safely, and we stood on the beach alone.

Alone.

We would not hear another outside voice for six weeks. The plane rose and quickly disappeared. We loaded the supplies in the boat and hung around in the sun for another hour. Jake and Jed wore calf-high rubber boots and spent their time throwing rocks and building mounds. I watched them closely. "Be careful going in search of adventure - it's ridiculously easy to find," wrote William Least Heat-Moon.

* * *

We had no way to contact anyone from the cabin or along the way on the rivers. It took us five years after our first family trip to discover **KIAM AM** radio. All of us were jubilant when we first heard voices at the homestead. It was like home we told ourselves. The signal at the cabin was weak but a linkage to the rest of the world. The first summer with radio, I climbed on top of the cabin and held the radio to my ear and relayed the news. Three times a day the station broadcast personal messages across the wilderness. People sitting in cabins, tents and boats across the interior of Alaska tuned in to hear if there was a message for them, providing a one-way link with civilization.

In town, I sought a more powerful AM radio. Hard to find. AM radios don't require strong reception capabilities because the signals aren't expected to travel far. The signals are weak on purpose so not to impinge on competing markets.

The next year I took in a radio that was supposed to pick up a clear signal, but it proved to be a disappointment. Another year passed. I took in a spool of copper wire. I wrapped the copper wire around the AM antenna, threaded the wire through a window edge and Jack ran it up a tree. Listening to the late edition of Mukluk Messages, broadcast at 9:15 p.m., with popcorn or Jell-O, became a special part of the day.

One time, in late July, a message came across the airwaves for me to call the school district office. To get to a phone, I had to run two hundred miles of river. We all talked about it, and James and I took off the next morning. After a night along the river bank and two hundred miles, I called. After about five minutes on the phone, I finished my business. It wasn't important. They just didn't know where we were or what was required of us to make a phone call.

* * *

We cruised slowly up the river. The airplane was gone. Jake sat on my lap and Jed trailed his hand in the water. Flax colored beaches reached across the river nearly closing it in places and bottom debris dominated the surface on some corners. I felt no hurry, just wanted to make it. On a beach, a few corners from the cabin, a black bear pawed at something in the sand. I cut the engine back and we watched. Thirty seconds later the bear headed for the brush. Its gait best described as a confident athlete trotting off the field. I accelerated ahead and around the corner.

We talked a lot about bears. Black bears crossed rivers, climbed on the cabin deck, and ran through the yard. They weigh over three hundred pounds and can run nearly thirty miles an hour. Agile, fast and powerful. It's estimated one hundred thousand black bears reside in Alaska. During summer months, black bears live in forest habitats, in low elevation river bottoms, and near drying river beds where there are lots of berries. Where we were.

I held the boat in the current in front of the cabin, so we could all take in the scene. The green roof of the cabin sitting one hundred feet back from the bank was visible as was a trail heading up the bank. I pulled the boat to a spot a few yards down river from the trail, and we unloaded before an audience of squirrels and mosquitoes.

We were tucked in bed that night when Jack heard a noise on the front deck and woke us. I saw the back of the bear through the front window. Jack grabbed the loaded rifle leaning in the corner, and we crept to the window. The bear jumped off the deck and disappeared. After several minutes, Jack, James, and I stepped out on the deck. A large black bear stood at the edge of the clearing in front of the cabin. It had a brown stripe from below its eyes to its nostrils. It stayed there for several minutes before sliding out of sight. I inspected the latch on the door.

We were all up early. I made breakfast, and we talked about bears and eventually moved to raptors. Eagle, hawks, ospreys and owls lived near the cabin. Eagles, were, perhaps the most often seen. I knew several bald eagles lived near the homestead. I'd seen them on every trip, their nests perched at the top of white spruce along the river. They soared above the river and frequently dove to the river and swamps. An adult bald eagle is more than three feet tall with a wingspan of seven feet. They have talons with a strength far exceeding a human's grip.

In shirt sleeves and under a brilliant sun that first day, we ran up river a few turns to fish for grayling. We drifted the middle of the river and took turns with two fly rods. This river, where it starts to rise quickly into the foothills, takes on an appearance of a mountain river, no boulders but widening and flowing much faster. There was room for a full cast, and I imagined us on a cover of a fishing magazine, the fly line curled in the air, and a boat full of kids. The grayling rose to everything we threw at them.

We all pointed to the sky. Not an eagle, but an osprey bolted below the trees and passed close to us, wings as wide as the boat. It glided and dropped, forcefully sinking its talons in the river. We all watched the osprey miss.

Raptors stop us no matter what we're doing.

Late that evening I saw an eagle while walking to the outhouse. I saw it on the ground near where the bank dropped to the boat. It sprung up and flew toward the river, the light from my lantern capturing it momentarily. About an hour later, with the gas lantern hissing softly as it cooled, we heard a thud on the roof. The boys whispered but mostly we laid still. Had an eagle stopped on our cabin roof? Maybe. It moved over the asphalt roofing for several minutes.

I loved that moment, then, and now. Take a handful of sand and throw it out onto the grassy yard. A grain of sand, the symbol of insignificance, landing among the green blades. Our cabin roof covered four hundred square feet in the vast Alaskan wilderness, and one of the world's most majestic birds stopped for a brief while. It can happen.

A few mornings later, Jack dashed into the cabin after a trip to the boat asking, "Is a porcupine edible?" He saw one up a tree near the river.

We talked about this some, okay. Jack shot the porcupine with a small caliber rifle, but it didn't fall out of the tree. He shimmied up twenty feet or so to dislodge the porcupine. Suspenseful moments as Jack inched closer to the porcupine and poked it with a stick to ensure it was dead. He prodded it out of the tree. The boys skinned the porcupine, and I boiled and fried it. We ate it all. Porcupines are herbivores, eating twigs and green plants. Mixed opinions on the flavor. I thought greasy beef. Not bad. We liked it okay, but one porcupine was enough.

Every morning we thoroughly discussed food. It went like this. During coffee and hot chocolate, Jed would ask, "What's for breakfast," and the conversation threaded through what we liked most. Jed told us what we had and how long it would last.

We ate fresh meat the first week of the trip, canned and dried foods thereafter. With fresh food gone, the main meals were spaghetti made with dried pepperoni sticks, potato soup, bean soup with canned ham, a rice dish with wild game such as grouse or rabbit, and macaroni and cheese with canned ham. Occasionally, we ate pike and grayling with canned vegetables or rice. Lunch was leftovers from the previous night or sardines and pilot crackers or peanut butter and jelly with crackers. Breakfast early in the trip was bacon, sausage and eggs with onions and pancakes, later in the trip, pancakes, oatmeal, and rice.

We collected drinking water off the roof. I installed a gutter on one side of the cabin and collected the water in a thirty-three-gallon plastic garbage can. A single heavy rain storm filled the container. We also drank the water from the river.

And there was candy. Lots of candy. We brought pounds of candy and split it equally six ways. Candy distribution was a ceremony. The agreement was each of us had received our total candy allocation for the entire trip. We were on our own after that. Jack, Jed and I devoured our candy and finished about midway through the trip. Jake made it to the last days, and Doylanne and James had candy for the trip home and held powerful positions.

Mornings we worked, sometimes in the yard clearing brush or constructing an out building, but most of time was spent on the cabin. We accomplished a lot of cabin work during this first trip, but many of the projects took years to complete. The first project was chinking the cracks and gaps between the logs. I used a rubberized substance about the consistency of thick, mashed potatoes to chink the cabin logs, brown in the interior and white on the exterior. A pastry bag and a cake decorating tip worked well, finishing with the backside of a teaspoon. We dipped the spoon in water and ran it across the chinking. It took thirty minutes to complete the length of one log, about forty hours to chink the cabin. The process was slow, but the chinking substance has retained its flexibility through the years.

Doylanne designed the interior. Soon knick-knacks, wall hangings, wooden pieces engraved by ants, river rocks, and wild flowers adorned the cabin. Book shelves were made with split birch logs resting on twelve-inch spikes partially driven into the logs. Mirrors on two diagonal walls reflected the logs from across the room, and on all four walls, Doylanne installed antique candle holders. Light moved along the contours of the logs with the changing brightness of the day.

I finished the floor with strips of two by twos on twenty-four-inch centers nailed to the floor, installing foam board insulation between the two by twos, and three-quarter inch plywood nailed on top.

Cabins are notorious for burning to the ground, so the wood stove was installed in the front right corner with thirty-six-inch clearance from the walls. The roof, where the stove pipe passes through, was built with sheet metal and angle iron rafters. I built a base for the stove consisting of a box with three inches of river sand. Doylanne tolerated scattered sand for a few years before we upgraded to a brick base.

Outside we cleared around the cabin, hauling brush to the river's edge, making a yard so the boys could play in view. Doylanne put bells on Jake and Jed's shoes to hear where they were. We hauled fire wood and carried logs, hiked ridges and climbed trees, hammered nails and made boards, dug holes and carried buckets, toted outboards and lugged gas cans. In the evenings, the boys pounded out two hundred pushups before bedtime. Wilderness boot camp. Our belts tightened and Doylanne and I each shed more than twenty pounds.

In the afternoons, we often played at a brown sugar colored beach several turns down river from the cabin. One afternoon a helicopter passed directly over us at tree level. It was flying so low we didn't hear it until it was right on us. The helicopter zipped over the beach but whipped around immediately. It hovered out there in front of us for a couple minutes like a giant dragonfly. The occupants undoubtedly enjoyed the scene of the kids, badminton net, and beach balls and certainly wondering what in the world we were doing. We waved and kept playing. It turned away quickly and disappeared. We must have been a strange sight in the wilderness.

We worked, we played, and we got filthy. Taking care of dirty clothing proved a special challenge. Greasy hands on pant legs, spilled fuel on a sleeve, fish slime everywhere. Grime multiplied, and the odor became nearly unbearable. Clothing could only be worn two or three days before it reached a disgusting condition where we all agreed it had to be laundered or ceremoniously sacrificed to the wood stove.

There is a rare and mystifying medical condition called trimethylaminuria. If you have trimethylaminuria, you smell like a fish. Imagine that malady. I've been there. The boys too. Rancid fish, reeking and rank, sour and rotten. We fished most every day and our clothing stank of fish slime.

With no running water or electricity, laundry took a full day with all of us involved. On sunny days, the clothing went to the beach where we laid in the sun, swam and did laundry. There were two classes of clothing, blue jeans, heavy shirts

and the rest. The jeans and heavy shirts were strung together with a rope and tied behind the boat, the rope run through the pant legs and through the shirt sleeves. The boat pulled the clothing through the water as a prewash. Up river and down. Clothes were then wrung out and put in the cooking pot heated on the campfire. Doylanne soaked the clothes in the pot for a while, stirring the heavy pieces with a stick before turning the rinsing over to the boys. The clothes went back behind the boat for a rinse. They ended up sort-of clean. The other clothing, the socks and underwear and t-shirts skipped the river and went straight to the pot.

On rainy or overcast days, Doylanne oversaw the work in the cabin. When this happened, a greater share of the burden fell on her, although the boys helped. She thought I should just stay out of it, although I was recruited to wring out the heavy clothing. The water was heated on the stove, and the clothing, including the filthy jeans, were soaked in the pot and wrung by hand.

At the end of our first trip most of the clothing was burned. Some we took home because we couldn't part with it. For years I wore a heavy, long-sleeve, blue and gold striped shirt. The boys called it the homestead shirt. Doylanne grew tired of washing it, and one morning it was presented to the jury for a final determination. Home or woodstove. In a rare moment in our family history, the boys sided with me and not their mother. The shirt was taken home and placed inside Doylanne's cedar chest beside her high school cheerleading outfit, our wedding pictures and a book of pressed flowers. Still at the cabin from the first trips are little tennis shoes, rubber boots, trucks, cars, wagons, and BB guns.

On cabin days, so cold and rainy outside the best option was to stay close by the wood stove, Doylanne read and the boys played cards, and I watched shrouded, snow covered mountains and a million drops of rain mark the river. It was during those moments, in this tiny, simple cabin, unique as a speck of flour dust on the miller's wheel, built on high, dry land with a lovely view of a small beaming river, our Alaskan wilderness dream became a reality.

One late July evening, following a week of scorching temperatures and mornings when the forest smoke smothered the river's surface, an extraordinary rainstorm pelted us. We were playing at the beach, and the clouds gathered so quickly there wasn't time to load all the toys. Shortly after getting back to the cabin, we heard the pounding drone from the blades of a helicopter. We rushed out onto the deck, and just above the trees the helicopter hovered. It held in place for a couple minutes before turning upstream. Minutes later the helicopter engine shut down.

Jack and I jumped in the boat. Three men stood on the bank about a mile up from the cabin. They were a federal interagency fire assessment crew checking a fire near our cabin, landing to get out of the storm. One crewman was from Alaska, the other two Montana. The pilot stayed with the helicopter. They were happy and quite surprised to see they weren't alone but declined dinner. Two hours later, the helicopter lifted off, the whine of the turbine shook the trees. We stood on the cabin deck and waited for the squirrels. Raindrops fell from the trees, and as soon as the sun broke loose, the squirrels began chirping.

On one gorgeous blue-sky day, we went on a river drive. We were out to see the sights. We ran down our river counting ducks to the first confluence and turned upstream onto another river. After twenty minutes running up stream, we saw a boat against the bank. I pulled in and shut down the motor. Immediately, we heard the playful screams of children coming toward us. Soon two small children and their parents were at our boat, surprised, of course, to see us.

They lived in southeast Alaska and were camping for a few weeks on land they staked during the 1980s land opening. First time we'd met them. We walked back to a clearing, and my eyes were drawn upward. Approximately twenty feet above the ground was a platform holding two dome tents. Below the Swiss Family Robinson tree-house, a circus net to catch high wire performers stretched among the trees.

They brought their children to the wilderness but were so afraid of bears, they built a home in the trees. Bells ringed the spruce holding the platform, and they retracted their ladder at night. They felt safe on the platform while they planned their cabin. I wouldn't let the boys climb to the platform. I feared them falling, even into a safety net. That couple impressed me. They gauged their fears and implemented a plan. We never saw them again.

We left the cabin for home on an early chilly morning in August in our cleanest clothes intending to run all the way to the launch. I circled in front of the cabin several times before pulling away. Doylanne wept, and we kept the cabin in view for as long as possible, but there was school, sports, and work to go to. We pulled away and were soon around the first bend. Bundled in the open boat with the certainty of a cold day ahead, we faced forward and watched an eagle make a wide turn overhead. Our journey was nearly over, but this six-week sojourn back in time shaped us like a clothing iron might, erasing wrinkles, defining seams, and pressing us in part, better or worse, into who we are today.

About 8:00 p.m., after more than ten hours in the boat, it was obvious we weren't going to reach the launch before dark. The wind was up and blowing strong against the current. Waves, breaking over the bow, soaked us. I had made a bad decision. What followed were harrowing moments that still wake me up in the night. We were in danger on the river that evening near dark. I'd forced it on a stormy day until the sun disappeared and the wind rose against the current. If the wind blows with the current the water is smooth, but against the current in a big river the water can become immediately frightening. I ran the boat on the edge near the sweepers. The rain pelted us cold like plunging an arm into the river. To run aground could swamp the boat, so I ran slowly in deep water next to the trees that reached into the river for us. The water began breaking over the bow. I heard a voice cry out from under the tarp. Doylanne watched me.

"Jack come here." It's always noisy in an open boat, particularly in a storm. But he heard me. He was next to me in seconds and we talked. A young boy but he got it. I told him to scramble up the bank and plant the anchor. He picked his way to the front of the boat, crouched and waited. A wave broke over the bow and drenched him. Doylanne screamed. I had my eyes on an opening in the trees to pull in and push up to the bank. Another breaker crashed over the bow. The boat was taking on water and everyone had come out from under the tarps. I turned in toward the shore, but too fast, and crashed against the bank, bounced back and forcefully drove into the bank again. Jack was gone. Up the bank more than ten feet he scrambled and disappeared; the rope snapped tight. We all watched. Jack was back at the edge in seconds smiling.

I sat for a moment as the boat settled against the bank and felt fright slip slightly away. "Stay put," I yelled. I moved to the bow. Doylanne followed me, and we guided each boy up the bank. We stood in the rain beneath the spruce, soaked but safe. I went back to the boat for a tent. The six of us slept in a two-person tent and were happy to do it. In the morning, the dome tent held the heat like a greenhouse.

At home after this first trip, I asked the boys what they missed the most while at the cabin. Easy answer: ice cream. We went again the next summer, and not knowing better, I dug a small hole through the active layer of soil to the permafrost and made a refrigerator. Every morning, Doylanne placed Jell-O in the hole with a jar of canned milk. In the evening, Doylanne added a whipped cream powder mix to the milk. One of the boys shook the mixture and was rewarded by getting to finish off the whipped cream jar and spoon. Not ice cream but close enough.

12

Irresponsibility, dads, and permafrost

We kept going, each year building on the year before. In 1987 the land was staked, '88, the logging was done, '89, the cabin built, '90, first trip with Jack and James, '90, first trip with Doylanne, '91, the first full family trip. In '92, '93, '94 and '95 we went as a family staying five to seven weeks each time. Every summer, I made other trips as well, some with Charles, some with friends, some with sons, some with Doylanne, some solo. There were summers when I made the trip four times.

The adventures became jumbled over time. We would talk amongst ourselves over the winter. "What year did that happen?"

In 1992, I chartered a plane to bring in a surveyor and take out Doylanne, Jed, and Jake. The staking rules required that I complete the survey that year. Jack, James and I stayed with the surveyor to help with the survey and to bring him out in the boat. The surveyor fit right in with us. I offered him a place to stay in the cabin, but he wanted to camp. He pitched a tent near the river. He surveyed the property in short order, and we helped by clearing the lines and carrying gear. Charles and I were only off a few feet when we staked the property with the hand-held compass, and the surveyor was impressed. The challenge we faced with the survey was that we needed to verify the township line. Two miles from the river monument was supposed to be another monument. Section and township lines in this area and through most of wilderness Alaska are not brushed.

We started working our way toward the monument, and after a half mile I asked, "If you had proof that there was a monument out there would that be good enough?"

"Yeah."

"The boys and I will go find it and bring you back a paper tracing."

"Sure," he said. "I'll sit right here." He was skeptical.

We took off with a compass. I held the compass on magnetic north, then made the proper adjustment for true north. Jack and James ran ahead of me and I positioned them with hand signals, then walked to them. We repeated this, and eventually using the topographical map to determine distance, found the monument. We triumphantly returned three hours later with a pencil tracing. We also discovered near the monument an amazing open meadow that we would use one day for an airstrip.

The run back to the boat launch with the surveyor was uneventful, but two memories make me smile. The first was the surveyor's astonishment at Jack and James throwing each other onto the sand. We camped going out on a beautiful pearl colored beach. The boys were wrestlers, and they practiced wrestling moves. Jack would throw James into the air and they would land in the sand, like tossing siblings flipping each other on a bed. Then it was James' turn. They knew what they were doing, and the impact was not nearly as violent as it appeared. That was new and strange to the surveyor.

The other memory was the surveyor's reaction when I pulled the plug out of the boat. We were cruising on a straight stretch when I reached back and pulled the plug to drain the water from the boat. He screamed when I showed him the plug. Who were these crazy people? I put the plug back in place and explained the procedure. He'd never seen it before. We all liked this guy.

***

One day in 2011, the year after my heart attack, while on a solo trip, I left the cabin with a backpack and a rifle to walk to the monument we traced for the surveyor. Slow going on purpose, I carried a tent and sleeping bag in case I wanted to stay in the woods. I had a satellite phone. It didn't much matter where I was, I figured. I crossed swamps and climbed the highest ridge in the area. I wove through rose bushes and stepped over bear and moose scat. I finally broke through the trees to a meadow the size of a small city. Miles of high russet grass swayed in front of me and I saw the Serengeti with its Sausage trees.

I picked my way across the meadow and sat on an incline and watched the Alaska Range and the hint of an airstrip that we all built more than fifteen years before in the mid-1990s. Charles, an excellent pilot, used that meadow for a couple summers. To make the strip, the boys and I fell a few small trees and used a gas-powered weed cutter on the chest high

grass along the edge of a dried lake. We carried and pushed a four-wheeler through brush to the meadow to drive back and forth to flatten the surface. One July, Charles flew hundreds of miles from Bethel to spend a few days with us before taking Jack out to a wrestling camp in Fairbanks. On that hot summer day, we stood on the edge of the airstrip when Charles lifted off with Jack. The plane climbed rapidly, brushing the top branches of a spruce. We all exhaled.

Another time, Doylanne and I were at the cabin when Charles and his wife Georgiana flew in to spend a few days. The plane swooped over the cabin, and I headed to the boat to run downriver to meet them. We had identified ahead of time the closest route from the grass strip to the river. I waited for them. More than an hour later I heard voices. I hollered back so they could get their bearings. The brush rattled with swear words and Charles broke through to the beach.

"Come this way, George," Charles called.

"Where are you, Charles? For heaven's sake." Georgiana pushed her way through the willows onto the beach. Her arms were full of bags of fresh vegetables. They were exhausted. The distance was over a mile, breaking trail. It was the last time we used that airstrip.

* * *

In 1995, Jed discovered a hole in the boat that led to one of our most frightening moments on the river. It was early June and we were heading to the homestead for a six week stay. The day was beautiful and the boat launch went smoothly. About thirty minutes into the trip, Jed said, "Dad, there's a hole in the boat!" He was squeezed between a box of food and a spare outboard motor near the bow. Doylanne took over the wheel, and I picked my way through our load to the front of the boat to look. There *was* a hole in the bow, just a couple inches above the river's surface, big enough to stick my boot through. I saw the river passing beneath us. I quickly wove my way to the wheel.

"We're going back," I told Doylanne. She pulled Jake onto her lap. The boys watched closely as I turned the boat carefully on one of Alaska's largest rivers, about a quarter mile wide at this point, while staying on-step. The boat had to stay on step when I turned against the current. It did. The boat slowed, but we were moving fast enough.

Half hour later I ran the boat onto the shore. We loaded all the gear back into the truck and trailered the boat the short distance into town. I stopped at a café and the locals directed me to a guy who welded aluminum. He worked on the boat, and I felt ill. The boys played near the truck while I laid back in the front seat. Doylanne made lunch, but I couldn't eat. I pulled the boat back to the launch and studied the river, tired and reeling with images. The boys ran down the beach. Alaska has the highest rate of drowning in the nation. Slightly more than 20% of those are occupational, the rest recreational. Thirty-six percent occur in rivers. Our river slithered and gurgled by, the water silver with silt, visibility below the surface nonexistent, the depth generally indistinguishable, the temperature about forty-five degrees. I felt sick.

We started again. The boys found their places in the boat and settled in for a long run. Doylanne's smile said it would be okay. On the river, I immediately began to feel better.

* * *

Safety. The goal was always safety, but it's so easy to get hurt. Before our very first trip with the boys, a deal was made: If they didn't get hurt, I'd pay them twenty-five dollars when we got home. State Fair money. They watched this closely, keeping track and frequently checking in with Mom. If one picked up a splinter, he'd run to Mom and see if he was still in the running for the money.

Handsome dark eyed Jake, only six-years-old, nearly lost his twenty-five dollars. I nailed a rectangular two by three feet piece of half inch plywood onto a stump for a small table in the cabin. The corners weren't rounded. Jake, eating at this table and standing in a folding lawn chair, buckled the chair and his head crashed against the corner of the plywood leaving a nasty cut to the bone on his beautiful little face. The gash, just below the left eye, streamed blood. Doylanne applied pressure to the cut, and I found the stitching kit. I held the kit in my hands for a while, opened the package and inspected the curved needle. I wasn't sure I could do it. This probably wasn't a good idea. Doylanne shook her head no and meant it. She held him while he sobbed. I couldn't do it. We used butterfly bandages.

I seriously considered running Jake to the hospital. Jack, on his own volition, carried the table out to the yard and rounded the corners with a rasp. The next morning Jake asked if he was still going to get his money. Today he carries with him a fading scar. He got his money.

Jack and James, too, came close to losing their twenty-five dollars. One day, they left the homestead for an upriver camping trip. Jack, fourteen, and James, twelve, were going about ten miles further upriver into some of the wildest, most picturesque country in the foothills of the Alaska Range. Doylanne and I stood at the river's edge, where a few lounge chairs were placed for sunbathing, and waved them on their way. The fifteen horsepower two-stroke and twelve-foot skiff shot onto step, and Jack, at the tiller, hit the first channel perfectly. It took a few minutes to run the straight stretch up river from the cabin before hitting the first turn. The trail of the outboard led to the boys and the sparkles fanned out behind them. They crossed the river in the channel, made the turn and disappeared.

They were headed to a beach across from the confluence of a little stream where we often fished. Their plan was to set up camp on the beach and fish in the main river or run across and fish the creek. That night they fished and cooked dinner on the gas cook stove and sat out at the campfire until late. Young boys. At bedtime, they put their supplies under a blue tarp and weighed the tarp down with a cooler and driftwood.

As they slept, their supplies burned.

When they awoke, all that was left was charred silverware and a stove frame and fuel canister. Clothing, life vests, rubber boots and all the rest burned. During the night a fire blazed only a few feet from their tent.

Doylanne and I didn't hear the outboard until it was on the final stretch to the cabin. Much earlier than expected, we were puzzled as we headed down to the river bank with our coffee to meet them. Their gear was missing. We huddled together at the edge of the river and analyzed this story.

How did this fire start? I think the gas canister leaked under the tarp and ignited. I don't know how. We now always disconnect those pressurized gas canisters after the stove has cooled.

* * *

I felt irresponsible at times when the boys were young. I often analyzed dangers at the homestead, using a process that involved justifying the exposure to scary situations by comparing them to experiences in town. "Just riding in a car is a lot more dangerous than this." "Can you imagine living in a big city?" "There are a lot of unhinged people in town."

Truthfully, though, I really didn't know.

One winter evening when Jed and Jake were in high school, we were talking around the dinner table about the changes to the land at the cabin and global warming. Why was the permafrost melting, not just at the cabin site but across the north? Opinions differed, but Jed had some news. "Oh, I know someone who has a place on the river on the way to the cabin."

We all looked to Jed.

"His name is Seth. You know him, Jake. Seth's Dad built a cabin on the river about twenty years ago."

Jed described the cabin and I knew exactly where it was on the river. Then Jed shared with us a poignant detail.

"When Seth was a baby, his Dad died."

Charles and I, on our trip to get the building materials to the cabin in 1989, discovered this cabin. We were loaded down heading upstream, rounding a corner, when Charles saw the glimmer of a metal roof in the woods. We turned back and hiked to a beautifully notched cabin. We walked around the building inspecting the craftsmanship. Seth's last name was engraved on a board above the door. I didn't recognize the name. Charles and I admired the cabin with its bear claw marks and pitch adorned spruce logs and hoped ours would be as nice. We stayed only a short time. We had a long way to go.

Several years later, I heard from Jed that Seth wanted to run a boat up the river to the cabin his Dad built, about a one-hundred-mile journey. He'd been to his cabin once before. His stepdad took him into the country by helicopter to see the cabin, but now he wanted to run the river himself. I called Seth and offered to run in there with him if he wanted company.

"I'm buying a used boat and motor," he told me. "It'd be great if you could go." Seth, about twenty-five years old, would prove to be a natural river navigator.

Both Seth and I ran four-stroke outboards, unlike what I ran in the early years. Almost all outboards were two-stroke when I started running rivers, so mixing two-cycle oil in gasoline was the norm, but they were nasty polluters. According to the EPA, two-stroke engines discharge as much as twenty-five percent of their fuel and oil unburned into the water. These smoky, polluting beasts are smelly too,

sometimes a lovely blend of rotten eggs, oil, gas, and rubber. They also emit far more hydrocarbon pollution than some much larger automobile motors.

We left from the launch on a warm June morning. He ran behind me in his eighteen-foot skiff. The water was low, so in places we picked our way through the river crossings. We camped a couple hours down from the launch, drank a couple beers, watched the undecided sky and discussed perseverance and hard work versus talent. I slept in my boat because I had a battery powered fan I wanted to try as a mosquito deterrent, and Seth set up a tent. The fan helped.

He started the morning fire and had coffee ready when I crawled out of my bag.

Four hours later we pulled up to a beach. We grabbed our rifles and hiked back to the cabin. "I don't remember this," Seth said, pointing to a boarded window on the river side of the cabin. He smiled, exhaled and his blue eyes studied the bear marks. "Let's go inside." The cabin was latched but not locked. He pulled the door open and we stepped into the dark. The floor was covered with fiberglass insulation pulled from the ceiling by squirrels and mice. "Let's get the boards off the windows so we can see," he said. We cleaned the cabin. Dirty but beautiful. A little bigger than the cabin I built. I would learn that the cabin was built in 1984.

I walked around looking for water and signs of melting permafrost. The land appeared in better shape than mine. The active layer looked sound. I was looking for soggy turf where it shouldn't be soggy. The active layer of ground cover, that which we walk on, is the few feet of soil and growth that freezes in the winter and thaws in the summer. Below the active layer can be permafrost.

Elizabeth Kolbert, in *Field Notes from a Catastrophe,* describes a surprising characteristic of permafrost:

> For the same reason it is sweaty in a coal mine--heat flux from the center of the earth--permafrost gets warmer the farther down you go. Under equilibrium conditions--which is to say, when the climate is stable--the warmest temperatures in the borehole will be found at the bottom and temperatures will decrease steadily as you go higher. In these circumstances, the lowest temperature will be found at the permafrost's surface.

The coldest part of the ice is on top, at our cabin about eighteen inches down. As the earth warms, the surface of the ice is more susceptible to melting. What's happening is the coldest side of the ice cube is being heated.

The land around Seth's cabin looked good, but there were drunken trees throughout the area, particularly closer to a meadow that began one hundred feet from the front door.

Seth found a journal on the table in the cabin. The first entry was dated ten years before. It looked like a frequent traveler on the river was using the cabin to spend nights occasionally. It'd been three years since he last stayed there.

Seth and I walked back to the boat for a cool breeze and to make dinner. We talked about the cabin and how he wanted to spend more time on the river. I got that. Seth would make his Dad proud. I stayed at the boat and Seth went to the cabin. He spent the night in the cabin his Dad built, and I anchored out in the river as far from the bugs I could get.

Seth and I cleaned the cabin for half a day before heading back to the launch. I ran aground once. I was daydreaming. Seth hung out behind me running his boat in circles, obviously enjoying the day. I saw him standing on a bench seat clearly in charge of the river. We made it home late that night. I've heard from Seth a few times since our trip. We're friends. He dreams about the wilderness.

* * *

Jack and I ran up the river in May 2016 to move the cabin out of the water. Under the cabin, the ground was pushing up against the floor, bowing the floor joists. Each year the water became deeper and the cabin sank. The soil, under the cabin floor, was sinking, but at a slower rate than the exposed active layer in the yard.

When the water first began to pool, I attempted to trench it off to the lower ground, helping temporarily, but within a

couple years the entire area was flooded. Early on I built a walkway we used for nearly twenty years.

During those years, I often used hydraulic jacks to level and crib the cabin with small logs so when we moved the cabin it rested on logs more than two feet above the original posts.

Facing the front of the cabin, most of the water was to the left and the front. The right was elevated some and there were a few large trees close, but water wrapped around the trees on the right. Knee deep water in places, the joke was one day we would fish from the deck.

2016 was the hottest year on record across the earth. Leaves sprouted in southcentral Alaska in mid-April, a full month ahead of their usual arrival. All my years in Alaska, I waited impatiently for leaves in mid-May. They would begin to show, tiny buds, and remarkably within a couple days be in full bloom. Mid-April leaves shocked us. Jack and I left in mid-May, and there were leaves all the way up the highway through the mountain pass.

The river trip went well, although we did have to install a new water pump on the way. We camped on the river and shot guns and talked. Jack, married with three children, was a physical therapist and had just opened a business in a different field with his brothers. There was a lot to talk about. I had retired from public education and was the director of a non-profit. We took turns solving each other's challenges.

We pulled up to the homestead the evening of the second day and were astounded at the depth of the water around the cabin. To move the cabin to the best reasonable spot, the highest ground within a doable distance without taking out more trees, required the cabin move to the left about twenty feet and forward for one hundred fifty feet, for a height gain of four feet or so, about a three percent rise.

We ran down river and found a stand of spruce. We cut three down, then limbed, bucked and winched them to the river. We pulled the six logs to the homestead and winched them onto the bank where we used a drawknife to debark them. We winched into the pond on the left of the cabin two logs to be used as rollers. On those we winched two of the largest diameter logs to use as runners for a sled. The cabin would ride on this sled the one hundred and fifty feet and the large logs would serve as the foundation at the new location.

The fun began. We planned this cabin move for years, and we thought we'd jack the cabin up until we could get the sled logs in place. That approach required we cut a few trees down to get to the new spot. Faced with losing trees, we dropped the idea. No trees were going to get cut down near the cabin and no holes dug.

We jacked the cabin up and slid a roller log under the left end of the foundation logs and winched the cabin to the left a few feet and onto the roller log. The cabin rested on the roller log and partially on the cribs. Teetering. The cabin had to go downhill now, a drop of at least two feet. We placed another roller log under the ends and winched another couple feet. The cabin was now on the two roller logs held by the winch cable and tilted at about a fifteen-degree angle. Another pull and the foundation logs should reach the first sled log. We hoped. They did, but the cabin angle was now closer to twenty-degrees.

Strange sight. We met out in front of the cabin and looked. Something was about to happen. We hoped something good. We took our time to get back to our places. I went back to the winch and pulled. I carefully accelerated, and the cabin moved and slid nicely onto the sled logs. The cabin now sat in a pond, the floor just above the water. Beautiful. The sled logs and the foundation logs provided just enough height. We would be there for three days.

As the cabin rolled upon the sled logs, the nearest runner log rolled a couple feet too far, placing it out of position for a sled. We jacked up the cabin and repositioned the sled logs.

We slept in the cabin that night. All looked good. We took stuff off the walls and disconnected the cooking stove, but the beds and wood stove stayed in place. Not straight, not level, not plumb, but okay.

Jack brought some metal plates to serve as a base for the jack. He painted them to resist rust. The jack we used, commonly called a farm jack, performed amazingly, receiving from me the most valuable tool award. We were able to jack the cabin up and pulled the sled logs in place because of this tool. Once the logs were in place, we positioned cross pieces in the front and rear and connected them with all-thread bolts two feet long. We cinched them up and were ready to move forward. Slower than we expected. We were in a good place, though. The cabin stayed in one piece. Now time to move the cabin forward.

The next morning, we used the same pulley configuration we had for the side pull. Three pulleys, one at the cabin and two at trees, and I was at another tree with the winch. I figured the cabin weighed about six tons. Some thought went into the calculation, but it was a rough guess. The winch was rated for three thousand pounds and the cable had three-thousand-pound tensile strength. Pulling the load straight without a pulley required a force of six tons. One pulley reversed the pulling direction but didn't reduce the force required. Two pulleys reduced the force needed to three tons, and three pulleys reduced the force required to three thousand pounds.

The winch drum held one hundred twenty feet of cable and I brought in another two hundred and fifty feet. After finding suitable trees, we ran the cable through the pulleys and crimped the cable ends onto metal loops. We were ready.

We stood in the yard surveying the layout when we heard noise near the wood shed. A black bear stood with its back to us, pawing the firewood. We watched for a time before shooing it off. It ran through the pond and over the mound where the cabin once sat, splashing through the water like it was running through surf.

I used a winch drum connected to a jazzed up 76.6 cc monster chainsaw. I asked Jack to get behind me. The winch rested on a fifty-five-gallon metal fuel drum anchored to a large spruce tree. I gunned it. Seconds later the cable broke about six inches from the drum, snapping like a whip toward the first pulley. I shut down and all was still. I saw spheres float before me. The whipping cable could cut a moose in half. The cabin hadn't moved.

We walked the length of the cable thinking out loud. I was confused. The short pull went well. We cut the cable and crimped on a new metal loop and restrung the pulleys. Something must have been in a bind. We pulled the slack out of the cable and inspected the pulleys to insure they would turn properly.

I tried again. I gave the saw gas and the line tightened and slowly more gas and the line snapped, breaking again near the winch. The cable sailed away from us. I was now spooked.

We added another pulley, the fourth. Neither of us had done anything quite like this before, so the going was methodical, thinking, questioning, walking through the motions. It was evening, time to quit, but we were ready to try again.

Don't break the cable! Slow. Easy on the throttle. That was me thinking and Jack talking.

If our calculations were correct, (our confidence was waning) with four pulleys, we were winching less than a thousand pounds. I gave the saw gas and held steady. Slightly more and held steady. There was a delay, but the cabin slid forward. I backed off and tried again. Same thing. The cabin was moving at about a quarter throttle. I shut down. It was time to stop.

Just after dinner, Jack shot a black bear, a beautiful six-footer. The bear hide now hangs on a wall in his home. Now we had much more work to do. The process began with gutting the bear with the entrails spilled onto a poly tarp pulled to the river and emptied. We didn't want bear guts near where we were working, always a chance bears and wolves would be attracted to the entrails. He skinned to the feet where only the last knuckle was left in the claw. He skinned the head next, splitting the nose and turning the lips, fleshing around the eye sockets and turning the ears. Turning the lips and ears essentially means turning them inside out and removing the tissue. To flesh the hide, Jack scraped meticulously then laid it out on a tarp and salted it, pouring fine salt on the hide and rubbing it in thoroughly. The next day, he salted it again. Jack rolled the hide tightly, hair out, for the journey home. Jack split the meat among family in town. The best bear meat I've ever eaten.

The next afternoon we were back at the winch. We pulled, and the cabin moved, but unfortunately slid right away against a root wad. We dug for half a day to make enough space for the cabin to pass. Later, near quitting time, under a clear sky and with the sound of trumpeter swans celebrating, the cabin surged triumphantly out of the water.

We stood behind the cabin and relived the trek of the cabin. We had pulled it out of our rising sea, but there was more work to do. Slow progress. Most of the time was spent re-rigging the pulleys as the cabin rolled up the incline. We did cut down one tree. The space simply became too narrow, so a beautiful birch became firewood.

Two days after the bear was shot, the cabin rested at its new home. The runners flush with dry ground, the original foundation logs on the runners, and the floor joists on the foundation logs. Sitting high. I picked the spot because it was doable, higher, and Doylanne wanted it.

I stood on the deck of the cabin and watched the river and thought of the fire-burned spruce standing thirty feet ahead. A couple years before, the cabin was saved by a crew of smokejumpers. A forest fire burned thousands of acres in the area, raging within yards of the cabin.

I watched the fire's progress that summer on the news and postings on the internet. I knew it was close to our property but didn't know how close. I had no idea there was an effort underway to save the cabin. The Alaska Interagency Wildland Fire Management Plan calls for evaluating safety and available resources to provide appropriate suppression actions on fires that threaten human life, identified private property, and high-value resources. We were fortunate that summer. In September, we went to the property and saw their work. The firefighters parachuted in and camped near the cabin. They cut a swath around the cabin and cleared brush and worked in tremendous heat and difficult conditions.

The cabin stood firmly but had nearly perished by fire and flood.

Before we left for home, Jack and I carried a chainsaw around the property line, clearing debris. I felt disappointed. The low spots were wet, some areas flooded. Leaning trees crossed overhead.

Jack and I saved the cabin in May, at least temporarily. Over subsequent years, we would fill the low spots in the cabin yard with rotting logs and leaves from the woods. In the large pond where the cabin once sat, we keep adding wood, advice from a retired permafrost scientist.

I learned that melted permafrost stabilizes and dries in thirty years or so. We may have already experienced the worst of the melt at the old cabin site. Now, perhaps, new soil made from rotting debris will absorb the water and eventually repair the area. One day the land may be dry again. It will be different—no permafrost—but it will be drier and warmer. The moose will move farther northward, and maybe elk and deer will have new habitat. I won't be around to see it, but my great grandchildren will likely love venison.

When I found the survey monument in 1987, and the land stretching toward Denali, I'd found the wilderness Edward Abbey wrote about in *Desert Solitude*. I've come to believe since then we all must find this place, whether wilderness parcel, back-yard garden, or urban park. Each of us, by finding a place worth protecting may help preserve the earth. Aldo Leopold wrote in *A Sand County Almanac:*

> Acts of creation are ordinarily reserved for gods and poets, but humbler folk may circumvent this restriction if they know how. To plant a pine, for example, one need be neither god nor poet; one need only own a shovel. By virtue of this curious loophole in the rules, any clodhopper may say: Let there be a tree - and there will be one.

The cabin was safe. Jack and I headed to town and back to work. He ran the boat, the water in the main river high and silver, thick with glacial silt, and I thought about twenty-nine years of river journeys. I was about Jack's age when I made my first trip. Inexperienced and dangerously naïve and reckless, I found a way up the river with Charles's help.

Through the years mistakes and mishaps have equaled triumphs and celebrations. My time stuck to a river bottom equals the time Charles and I spent building the cabin. I've known this little cabin half my life; my hair is now the color of the silver river.

* * *

In 1996, Jack was heading into his senior year in high school. We hadn't made the move to the homestead. Everyone was neck-deep in something in town. Life was kids, school, and work. Up at six, work all day, all the duties of working people. Pay the bills. Get ready for winter. The kids did well in school. One day in April, a few years after that first full-family trip, I came home from work with news: "I got offered a new principal's job." I'd just spent an hour talking with the superintendent about results-based accountability and school reform. He offered me a bigger school, more staff, more work, longer contract, and more money.

"Are we going to the cabin this year?" Doylanne asked. We both sat quietly. Life had certainly changed. Work, school, sports, and friends, all pulled at us. In ten years of wilderness travel, my wilderness dreams became foggy like a mirror after a long shower. When I wiped my hand across the glass pane, I could see a distorted image looking back at me. W. G. Sebald wrote in *Rings of Saturn*: "It takes just one awful second, I often think, and an entire epoch passes." It was only a second. Wilderness living became wilderness experience. Escaping the craziness of the world became seeking a balance. Concern about only the necessity of wilderness became concern about conservation, our local environment, and protecting the special places we love.

"Yeah, we're going," I said. I was here and there, living in both places, connected to both. Our river in Alaska and all of us and every creature share the earth in the present and in the past. We are connected whether we want to be or not.

That next summer, Jake, with a nice scar near his eye, nine-years old, ran up to the homestead with me for a two-week stay. We ran into some problems on the trip home. The fifty horsepower two-stroke ran well, but the lower unit took a beating from the river bottom. After a particularly hard smack, the forward gear quit working. Reverse worked, but forward wouldn't engage. I tried to run the boat downstream backwards. Bad idea, but it didn't matter much because now I couldn't get reverse to work either.

I poled the boat over to the nearest beach. We weren't going anywhere for a while, so we erected the tent and built a campfire. Jake played on the beach. He didn't mind the extra night on the river. He chased a butterfly until I hollered at him to come back to the camp. Most of the time, though, he stood by, fetched tools and helped me with the outboard.

I tinkered with the shifting linkage for a while before pulling the transom up onto the sand to examine the lower unit. The problem was obvious. The bolts meant to hold the lower unit onto the housing were missing causing the lower unit to sag to the point it disengaged with the motor. Quite unusual, and a mystery to this day. Pretty rare for all four of the bolts to vibrate loose and fall from the lower unit. We had no way to fix the motor. The bolts were gone.

Jake and I pulled the stern as far as possible onto the beach but lacked the strength to pull it entirely out of the water. After disconnecting the steering, power and throttle cables, we lifted the outboard off the transom. I needed him. The fifty-outboard weighed more than two hundred awkward pounds. He struggled, so did I, and we finally laid the outboard in the front of the boat. We installed the spare thirty horsepower two-stroke on the transom and were ready to leave in the morning. We awoke to fog and headed downstream but didn't have the power to get on-step. I rearranged the load several times, but it wasn't happening. We ran to shore and pulled the motor into the brush to be brought out another trip. The boat propelled by the thirty horsepower finally rose on step. Jake was bundled with all I had for him, and he crawled under a tarp. I had on three shirts, an old letterman's jacket, with the last layer a life vest. A knit hat covered my head and I wore cotton work gloves. I stacked a couple bags of gear on top of a fifteen-gallon fuel drum in front of me. I warmed my cheeks with the palms of my hands.

After a couple hours on the river, water covered my ankles. The rain filled the boat. On the next long stretch of river, I pulled the drain plug. The plug hole, a couple inches above the river surface with the boat on step, allowed the water to drain under power. I watched the water level drop in the boat and the oncoming corner approach. It would be disastrous to lose power or run aground. The water drained slowly. I crossed the river at the corner as the water at the drain hole began to whirlpool, signaling an end to this ten-minute adventure. I scooped out mud and leaves with my fingers, and as fast as I could, replaced the plug.

On one long straight stretch, I saw some activity near the bank. I stopped the motor and drifted toward a large beaver humped over with a birch branch in its mouth, another beaver watching. The beaver partially hidden in birch leaves, slid into the water pulling the branch downstream before slapping its tail and diving.

I wasn't seeing too well either. Four days before, while hiking behind the cabin, I laid my glasses down while taking a break. I left the russet frames on the mostly brown forest floor. I searched for more than an hour, disappointed in my carelessness. I never found them. The silver river with clouds resting on the water made a gray world. My eyes hurt but I couldn't close them. The water passed by too rapidly. I watched the debris, bubbles and ripples through squinting eyes.

The rain stopped in early evening. Tall white spruce with dark needles mixed with birch with emerald leaves lined banks for miles and miles. As the clouds lifted, I was drawn to a dark streak in the sky just ahead dissolving to pale blue. I stared like a child searching for the center of a marble. Upriver the painted sky began white at the horizon and intensified to red overhead, and the river faded from silver in the ripples to deep green near the sweepers. Before me was the most beautiful sky I had ever seen.

There are moments when we all like the cold. I sat up in my seat smiling at the wind, lifting the knit hat above my ears. Soon we were at the launch. We sat in the pickup waiting for the heater to get us where the chill was gone. I remember that moment clearly. I listened to the truck motor and the heater blower. Little Jake shivered and smiled. The change from cold to warm in the truck, like crawling under blankets or stepping into the shower, relaxed me, and I settled back in the seat. Hermann Hesse wrote in *Siddhartha*:

> Have you also learned that secret from the river; that there is no such thing as time? That the river is everywhere at the same time, at the source and at the mouth, at the waterfall, at the ferry, at the current, in the ocean and in the mountains, everywhere and that the present only exists for it, not the shadow of the past nor the shadow of the future.

"School will start soon, Jake." Are you ready for that?"
"Yeah." He pulled on dry socks. "I'll see my friends."

I drove the truck to the edge of the launch, rolled down the window, and shut down the engine to listen to the river. It gurgled and sighed. Nothing I know sounds like a river, except the cry of a newborn, the freshest sound of life. We hear running water naturally because we hear ourselves. We are a river, 100,000 miles of arteries and veins, little rivers of blood carrying oxygen. All the flowing streams, large and small, fast and slow, narrow and wide are necessary for the whole.

That was it. I was ready, finally, to accept that's it's okay to adapt a vision. After years of planning for full-time wilderness living, I pulled the plug on the dream that just didn't fit what we all needed. The river and cabin would do just fine without us. We didn't have to live full time at the homestead to experience wilderness and reap the multitude of benefits offered by nature. There were many things more important than my dream of wilderness living. One of those shivered beside me.

During the five-hour drive home from the launch, following the RVs and watching for moose, with Jake sleeping, a dream faded softly and quite gently, away.

Epilogue

Alaska and wilderness

Doylanne and I recently met a couple on their first trip to Alaska. We sat near them in a restaurant in Anchorage, and they told us of the wonderful visit to Alaska they were experiencing. They were catching a flight back to Denver, and they were reliving their trips to Denali National Park, Whittier, and Homer. All spectacular, but there was one moment above all the others. In their rental car, they drove to a small lake off Maud Road just out of Palmer, an hour's drive from Anchorage. They were sharing a picnic lunch on the shore of a pristine mountain lake when they were overcome with an uncommon sensation. It dawned on them—and they were astonished-- they were alone. The lake was theirs. They hiked a bit down the shore, immediately more cognizant of bear sign. For the first time in their lives, they said, they felt the wilderness. They had found wilderness Alaska on their first trip to the north.

Wilderness in Alaska is not hard to find. Fifteen of the twenty largest wilderness areas in the United States are in Alaska (our place is in one of those), so remote country is easy to find for those who venture out a bit farther than the few cities across the vast state, but what is surprising to many is that with nominal effort, wilderness can be found much closer to home than our place in the woods. Sadly, many people who move to Alaska are gravely disappointed because they can't find a place where solitude and natural beauty merge because of our accepted definition of wilderness. Newcomers are quickly led to believe the wilderness can only be reached in a hundred-thousand-dollar float plane or a forty-thousand-dollar jet boat, or the step down in price option, the chartered float plane ride at hundreds of dollars per hour. When they go fishing at the hotspots on the Kenai River, Russian River, or Deep Creek, for instance, they face large crowds and experience the astonishment of combat fishing. The same experience is found in Anchorage on Ship Creek and north to the Matanuska-Susitna Valley at the Little Su River, Deshka River, Willow Creek, Little Willow Creek, and Montana Creek. It's combat fishing, a battleground of flying pixies and t-spoon lures, drunks, pistols, beer, and toilet paper. Many people like this. Some love it. It's a colorful show, great fun, and intensely competitive. Some can turn combat fishing on Montana Creek into a trip to Katmai National Park. It's a mind game, and I tip my hat to them. A special art.

It's easy, though, to become quickly disappointed in Alaska if you've come here for wilderness experiences. It's easy to love this place, and just as easy want to give up. What to do? Some people have found the answer. They resolve to go find wilderness on their own.

Here is an easy way to find wilderness in Southcentral Alaska. Drive north of Anchorage beyond Wasilla to the Nancy Lake Parkway, sixty-five easy miles. Turn onto the parkway and drive a few miles. Take a compass bearing due north (or south) and walk a little bit. If you come to a pond walk around it. If you come to a stream, wade across it. Sit and relax. You have reached the wilderness. Alaska is full of these places, many close to the urban world.

No one will be stopping by to check on you. It's yours for a sliver of time. If you like it, you can come back, and at least for now, you will most likely be alone. One warning, don't get lost. Mosquitoes will devour you if you are not prepared, and you will eventually want to find your car.

This strategy would work in many other places too. Doylanne and I drove east of Salem, Oregon, to a park in the Cascade Mountains not long ago. Beautiful spot with well-maintained paved parking and a groomed trail leading to impressive waterfalls. We walked to the falls and enjoyed a stunning place, but my thoughts were back in the trees. Next time, we will take a compass and walk a half mile off the path and touch the Oregon wilderness.

Wait. I'm not supposed to do that. In many parks, maybe most, going off trail is not allowed or is at least discouraged. Hikers will encounter a sign reading something like: "Please stay on the trail to preserve the natural environment beyond the trail." I understand the rationale, but I believe the reasoning is short sighted. Restricting access protects a place over the short term but does not positively impact attitudes toward the environment. Unfortunately, people don't much care about what they don't know about. Walking down a groomed trail is fun and beautiful, but there is so much more beyond the trail. Nothing new here, I know. Again, wilderness is a necessity, as Edward Abbey wrote.

Wilderness is not a bear, moose, caribou or fish. That definition is the handiwork of the marketing world. This book has been about the kind of wilderness where Jedediah Smith trapped beavers and fought grizzlies. Some still exists across the world, but it's dwindling fast. The wilderness *idea*, though, is alive and well, perpetuated by big business and promoted for wealth. Huge industries exploit our fading memories to sell products to take us part way to nirvana as we combat fish on Alaska's southcentral rivers or hunt for elk on a hillside with five other hunters in Eastern Oregon, all of whom, in Alaska and Oregon, will retreat to recreational vehicles for the night. Automobile companies love the wilderness. I marvel at the truck commercials, always in the wilds pulling logs and crashing over mountainous terrain. Outdoor clothing and gear have expanded far beyond utilitarian need. At Sheep Creek in Anchorage, maybe the most urban salmon stream in the world, fishermen catch salmon in two hundred-dollar chest waders with three hundred-dollar rods and reels. The Ted Stevens International Airport during the summer and fall is a runway for the finest camouflage clothing available. Selling wilderness is obviously big money. I've bought a lot of that outdoor gear myself through the years. I admit, it's part of the fun, but it's not all necessary to experience wilderness.

The boys are taking their own trips to the homestead, and one day they will likely take their children there, and I hope they're all dressed warmly. They will have valuable skills and a few tricks, but they'll never understand most of it, any more than I do. I started out embarrassingly green, and I'm still a ripening banana on the window sill. I have learned, though, a few lessons: A river (all the natural world) is far too complex to fathom. Aldo Leopold wrote in *A Sand County Almanac*, "No matter how intently one studies the hundred little dramas of the woods and meadows, one can never learn all the salient facts about any one of them." Rivers are also unforgiving. When the day is radiant, and the water is high, there is a sensation of invincibility, but this is always temporary. Something will happen on a wild river to take your breath away and demand attention.

Where do I go to now? The homestead sits where there are no roads and no reliable communications, still. Technology has advanced to where some communications are possible with the right equipment, but it might not do much good to call emergency services. I hope I have successfully described how hard it is to get there. It's wild. Predators sometimes track down and kill moose, and they also will occasionally track down and kill a human. What's to come of all the effort to push toward the mountains? Will the next generation share the enthusiasm, and will other generations to come? Staking and developing the homestead was spurred on by boyhood dreams and a desire to see something through to the end. It grew to much more. Now in my mid-sixties, I wonder how many more of these trips I can take? They weren't easy thirty years ago. I buy things simply to make it more pleasant, something I never did before. The best heaters, the best lanterns, the best sleeping pads and on and on. A moose must jump out and scream shoot me and fall in the boat. But the trip on the river is the same and won't soften for me. I still run aground and repair outboards and chop firewood and mill boards.

One time when the kids were little, we stopped at a log cabin on the river. The owner had made a beautiful piece of wood work as a monument to his Dad. Friendly guy, he showed us around his place. A few years later, I ran into him on a cold day on the river when the water looked painful. His elderly mother was in the open boat traveling to his place. She sat upright in the boat in a chair, wrapped in white. It was both eerie and beautifully moving. That image has stayed with me for years. I hope one day I'm taken into the country on the river when I probably shouldn't go. I'm healthier coming out from the cabin than going in. I get on the river out of shape, heavy, and soft and come out tighter and stronger. We still carry water from the river to the cabin. Early in the trip, I struggle to haul a five-gallon bucket of water, but by the end of the trip I carry two. I would live longer if I lived out there, if it didn't kill me.

Find the place you love.

Acknowledgment

This book was abandoned, then revived, then neglected, then revisited several times before it became something I wanted to share. Two friends, Megan Moore and Jeff Ramseyer, looked at the manuscript at this early stage and offered valuable suggestions. Jeff read and commented on three versions, and Megan offered a thorough and insightful evaluation of the early manuscript. My son, James, the author of the Foreword, and books of his own, provided personal writing instruction, encouragement, and many ideas that shaped this book. Charles Rangeley-Wilson shared his thoughts about the narrative and suggested books to read to improve the structure. Gene Robinson offered suggestions on transitions and structure. Sharon, Marilyn, Charlotte, Delores, and Charles, my siblings, were encouraging, and of course, it just wouldn't have happened without Doylanne and our sons' support. Thank you all.

About the Author

You can't have too many boats or guitars; that's what Eric Wade says. Of course, an outlook like that won't get you on the fast track to wealth and success. He says that too, but apparently, doesn't care that much. He'd rather be on a fast track to the backwoods where he can mill white spruce boards and chase animals through the trees. He grew up in the Oregon coastal mountains where leisure time was spent stalking trout and work was always related to a log. After moving to Alaska, when not pursuing his dream of wilderness living, he taught English, ran schools, and led nonprofit corporations. A 40-year Alaska resident, he's keenly interested in the education of special needs students, conservation, outdoor education, and the intrinsic value of wilderness. He has a Master's Degree in Journalism from the University of Oregon. He and his wife, Doylanne, live in the Matanuska-Susitna Valley in southcentral Alaska.

www.ericnolanwade.com

Lightning Source UK Ltd.
Milton Keynes UK
UKHW011813181119
353758UK00007B/1991/P